ORIGIN OF OCEAN

RECENT ADVANCES IN ENVIRONMENTAL ECOLOGY SERIES—1

Origin of Ocean

A. P. DIWAN
D. K. ARORA

ANMOL PUBLICATIONS PVT LTD
NEW DELHI-110 002

ANMOL PUBLICATIONS PVT LTD
4374/4B, Ansari Road
Darya Ganj, New Delhi-110 002

First Edition, 1995

ISBN 81-7488-110-7

Printed in India

Published by J. L. Kumar for Anmol Publications Pvt. Ltd., New Delhi and Printed at Amit Enterprises, 518/14, Anup Mkt., Maujpur, Delhi.

Preface

The oceans are not equally distributed over the earth. Oceans cover more than 80 per cent of the Southern Hemisphere but only 61 per cent of the Northern Hemisphere, where most of the earth's land masses occur.

Whereas all the major oceans are connected to each other, the world ocean has been separated for convenience into four major divisions; Pacific, Atlantic, Indian, and Arctic, in order of decreasing size. Projecting from or partially cut off from, these larger oceans are smaller marginal seas such as the Mediterranean, Caribbean, Baltic, Bering, South China, and Okhotsk. The Pacific, Atlantic, and Indian Oceans all converge in the area around the Antarctic continent, giving a contiguous body of water around the southern continent.

This book deals with the chapters Introduction, Origin of Earth, The Climatic Environment, Freshwater Environment, Sedimentary Environment. Deep-Sea Environment, Island Environment, Active Animals of the Deep-Sea Floor.

AUTHORS

Contents

1

Introduction

The marine and terrestrial environments differ in several respcts. The most obvious difference relates to the supporting medium and its relative stability. The medium of the marine environment is liquid and in constant motion. Water transports both physical properties and both from one place to another. As a result, short lived plants dominate in the marine environment. These plants have evolved to cope with constantly changing physical conditions of the ocean. In contrast the terresrial environment has a soil which has provided plants with a far more stable medium. The stability is due to the fact that physical properties and geographic positions persist over long periods of time. The terrestrial environment has promoted the evolution of long-lived plants with specialized multicellular appendages such as root systems and above ground structures. They also have adaptive features for the particular physical conditions of the place where they reside.

The temporal and structural nature of the organisms of the two environments is directly related to the type of organic compound which prevails in the sea and on the land. Proteins are the dominant compound of life in the seas and carbohydrates are the major compound of land life. The type of organic compound dictates the basic operating differences of the two environments. The protein-based life of the seas is capable of growing rapidly but it does not have an ability to store energy for long periods of time. On the other hand the carbohydate based life of the land is relatively slow growing but able to store energy for extended periods of time.

There are several ways in which the physical properties of the marine and terrestrial environments influence the development, structure and styles of existence of organisms. For example, the influence of gravity plays a significantly different role in the lives of organisms of the sea and land. In the sea, pelagic organisms exist within an amniotic-like fluid which supports and sustains life. The fluid buoys the mass of the organism and is the source of the required minerals. In contrast, terrestrial organism must also have a structural framework that supports both the weight of the fluid and of the structure itself. This structure must be able to withstand the rigorous strain of the pull of gravity. Furthermore, the structure requires an added expenditure of energy that is borne by terrestrial organisms for development and maintenance.

As a result of the different environments, the energy requirements for locomotion are distinctly different for the organisms of the marine and terrestrial environments. In the seas, many organisms move with the fluid and require little or no energy. Even in active vertical or horizonial movement, much less energy is required by sea organisms than by land organisms. On land, each step that an animal takes requires the lift of both the mass of the structure and the mass of the self-contained fluid against the force of gravity. The majority of non-marine animals have adopted flight as a means of locomotion and have thus confronted the difficulties of terrestrial travel.

Besides influencing energy expenditures by organisms of marine and terrestrial environments, gravity also plays another role : it influences the nutrient status of each environment. Nutrients do not accumulate in the surface waters of the ocean. Instead due to gravity they tend to become depleted from the surface waters and sink to the ocean bottom. It has been estimated that without mixing and oceanic circulation primary productivity of the marine environment would cease in less than a year due to the depletion of the nurient supply at the ocean surface. In contrast the soil of many terrestrial sites tend to accumulae nutrients. During the life cycle of terrestrial plants, some of the production undegoes decay and becomes part of the soil medium. Over time an accumulation of nutrients may occur which in turn enhances the stability of the terrestrial environment. In essence the marine and terrestrial organisms and their supporting environments differ markedly from each other. The next two sections provide additional details on the physical basis of the biogeographic patterns of organisms of the two major environments of the Earth.

Physical properties of the marine environment

The physical properties of water, such as high heat capacity, latent heat of fusion, thermal conductivity, and dissolving power are characteristics which are important to organisms. Water is densest at 39°F (4°C). Becoming lighter as it freezes, ice form at the surface of water and insulates the underlying water forms continued rapid cooling that results from evaporation. The Earth would not be a moderately warm environment if water behaved like other substances.

The temperature extrems of the ocean are relatively narrow compared to the extremes within terrestrial environments. Temperatures range from about 28°F (−2°C) at the ocean bottom near Antarctica and the Norwegian Sea to about 86 F (30°C) in shallow waters at the Equator (Smith 1976). Also, due to its salinity, ocean water does not freeze as readily as freshwater lakes. The freezing point and temperature of maximum density of sea water both decrease with increasing salinity. At the normal salinity of 32-38 parts per thousand; vertical convection continues until the whole body of water has been cooled to the lower freezing point. This extra cooling requirement causes a lag in the formation of marine ice compared with freshwater ice. This phenomenon contributes to the climatic dynamics operating between marine and terrestrial environments.

The relative proportion of elements that make up the major constituents of 99.9% of the sea water is nearly constant. Because of the constancy of elements it is possible to measure the salinity of sea water by measuring any single constituent of the sea. The total quantity of dissolved salts, or **salinity**, is variable from one place to another. Salinity used to be calculated by measuring the quantity of chlorine, but now electrical conductivity is the most commonly used measurement of salinity.

In addition to the major constituents, sea water contains virtually every other natural element, but they are in small concentrations and highly variable from one place to another. The surface waters. where photosynthesis is most active, tend to be depleted of dissolved nutrient elements; while deep waters are enriched in these elements through settling of detrital matter (Smith 1976). In view of the limited nutrients at the surface, mixing and circulation of ocean water become critical to the understanding of the biogeochemical system which supports organisms of the sea.

There are may types of perturbations of sea water, including internal movements which resemble the massive hurricances or tomadoes of the atmosphere. One source of these movements is the Gulf Stream. As it meanders along the east coast of North America large loops of water break off and form either warm-or coldcore "rings", or massive eddies of water which move across the open ocean. There are about ten cold-core rings and three warm-core rings at any one time in the North Atlantic Ocean. These eddies of water extend from the sea surface to a great depth in the ocean. cold-core rings last two to three years while warm-core rings last about a year. Rings have been recorded as coalescing and breaking up. Eddies rtngs and of water have been descrtbed in many places and apparently can transfer large pools of water over long distances (Kerr 1977, McDowell & Rossby 1978). The rings and eddies of water play an important dynamic role in mixing and creating variety in what might otherwise be stagnant marine waters such as the Sargasso Sea (Backus *et. al.* 1981)

Marine waters of different temperature and salinity intermingle and thereby help to mix the marine waters (Kerr 1981). The intermingling has been likened to the meeting of fingers in a hand clasp and referred to as **salt fingers**. Atmos-

Table 1.1
The principal constituent elements of sea water (after Smith 1976)

Ion	*g/kg of water of salinity 35%*
Chloride	19.353
Sodium	10.762
Sulphate	2.709
Magnesium	1.293
Calcium	0.411
Potassium	0.399
Bicarbonate	0.142
Bromide	00.673
Strontium	00.079
Boron	000.445
Fluoride	000.128

pheric storms apparently churn up the near-bottom cloudy layer in the deep ocean. The ocean has many processes which operate to stir and mix water and sediments. These types of findings counter the notion gained from casual observations at the surface that the ocean is mostly a large and not very dynamic pool of water.

The distribution pattern of marine organisms is closely intertwined with the physical properties and dynamics of water. Several properties of water pose paradoxes for marine organisms For example, heat and dissolved oxygen and carbon dioxide are important in growth processes. However, the amount of gas absorbed by water is inversely related to its temperature. For that reason, warm tropical waters tend to have a low level absorbed gases, while the cold waters of high latitudes tend to have high quantities of absorbed gases. Tropical waters are suitably warm for optimum growth but their low levels of absorbed gases, such as oxygen, limit the number of organisms that can be supported. For example, the level of oxygen drops from a supersaturated 5 m/l to a nearly anoxic 0.2 m/l within the top 150 m of tropical waters. At high latitudes, plants are not as limited by dissolved gases as they are by low temperatures.

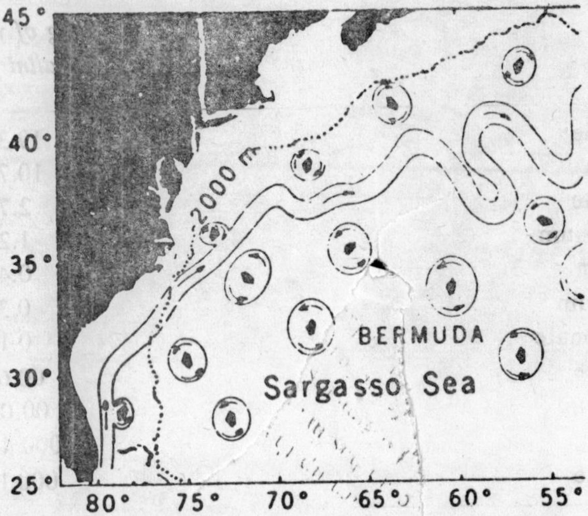

1.1. Ring eddies of the ocean are the ocean equivalent of hurricanes and tornadoes of the atmosphere.

A second property which poses paradoxes for marine organisms is that photosynthesis can only take place within a narrow range of light intensity. Light intensity in water is limited by absorption and turbidity. Water absorbs light energy and stores it as heat. The major factors which influence turbidity are terrestrial runoff, bottom mixing on the continental shelf, and the quantity of living and dead plankton, or **seton.** Table 1.2 compares the light penetration of sea water at nearshore and offshore sites. Note that the penetration of light into water rapidly decreases in the top 20 m. The nearshore areas have the greatest potential for plant growth because of mineral availability, however the increased seton contributes to the reduction of the penetration of light within these same waters. Also, day and night periods differ with latitude. In the Tropics there is a diurnal pattern in the light, while there is a seasonal gap in light within Arctic and Antarctic latitudes.

Table 1.2

Comparison of light penetration depth (m) of sea water at nearshore and offshore sites

%Light absorption	*Nearshore*	*Offshore (16 km)*
50	2	4.5
75	45	9.0
90	8	17.0

Finally, as described above, nutrients at the surface of the water are limited because of gravity. Organisms absorb nutrients, die and sink. The loss of required nutrients, especially available phosphorus and nitrogen has been found to severely limit the growth of marine phytoplankton. Successive layers of organisms live on the organic detritus as it settles toward the sea floor. Eventually, the materials arrive at the sea floor where some elements may be decomposed by microorganisms and become available for reentry into the marine system.

Only a few regions of the marine environment suport a large biomass. Meteorological conditions the dynamics of the atmosphere-ocean system, and the geographic arrangement of mountain ranges and coasts as dictate most of those regions. Four regions are particularly productive. Within the middle lattitude area of the Earth, coastal estuaries are exceptionally productive. As part of the interplay of polar and tropical air masses, the seasonal flooding cycle operates to transport, via stream systems, substantial quantities of sediments organic matter from the floodplains to the estuary. Mid-latitude estuaries undergo wide swing in productivity because of seasonal temperature fluctuations and variations of the amount of runoff.

A second type of condition that dictates high productivity is turbulent regions of the ocean where warm and cold currrents meet. One such place is the Grand Banks. This is the site of turbulence created by the collision of the Gulf Stream and the Labrador Cold Current. The turbulence brings nutrient-rich waters from the sea floor to the surface and supports a locally large biomass.

A third phenomenon that creates a productive region is where upwelling of cold nutrient-rich waters occurs along the west coast of continents as a result of the Ekman Spiral. These areas are subject to large wings in productivity as a result of El Nino-Southern Oscillation phenomena. At irregular intervals that vary from two to ten years the Trade Winds are unusually weak over the tropical Pacific Ocean. As a result, upwelling becomes disrupted. When up welling ceases, nutrient sources are no longer available and the growth of marine plants virtually ceases, causing devastation to the biological systems of the coastal zones of Ecuador and Peru. In the 1982-3 El Nino, chaos occurred within extensive regions where organisms depended upon upwelling. But, in addition to a collapse of the fisheries of Ecuador and Peru whole bird populations were reported as abandoning *en masse* Christmas Island in the equatorial Pacific.

A fourth, extremely productive region in the marine environment, is the polar seas. During polar summers, glacial melt water enters into the high-latitude oceans. Melt water carries large quantities of nitrates and silicates in suspension which are a major limiting nutrient for the algae inhabiting these regions (Apollonio 1973). In the polar environment the biomass undergoes a large seasonal fluctuation.

Only a small percentage of the marine environment has the suitable combination of physical conditions to support a large biomass. Figure 1.2 illustrates the distribution pattern of production of biomass found in the marine emvironment. Productivity is based on the growth of single-celled algae that cannot store large reserves of carbohydrates. Instead they rely on rapid reproduction and growth to replenish and maintain their mass.

Interaction of marine and terrestrial systems

None of the Earth's system stands alone. All systems interact with each other to make a longer, more elaborate system. The marine and terrestrial environments illustrate this point. The hydrologic cycle and global circulation patterns of the Earth provide a rational basis for an overview of the interaction of the two system.

From the point of view of the hydrologic cycle, and the movement of organic matter and minerals, the terrestrial environment is essential to the productiviy of the marine environment. The marine environment supplies water and the evaporative surface necessary to maintain the hydrologic cycle. The land is a major recipient and benefactor of the water but ultimately the water returns to the ocean. As it returns through channels of branching streams, water carries a load of minerals and organic materials which provide the nutrient base for many organisms of the marine environment. In this way we can understand that productivity of the marine environment has its roots in the terrestrial environment. An exchange is also found which influences the organism of the terrestrial environment.

The global circulation pattern involves the movement of large volumes of surface water toward the Tropics. This was as part of the Ekman Spiral and Coriolis Effect. The returning water becomes organized into an extremely persistent current. These currents strongly influence the climates of the South-eastern coasts of continents of the Northern Hemisphere and the northeastern coasts of continents of the Southern Hemisphere. They also influence the distribution of organisms of the Earth. The Gulf Stream, also known as the Floreda Current, illustrates the influence which these phenomena have on the organisms of the Earth.

The first good chart of the Gulf Stream was made by Benjamin Franklin and Timothy Folger and printed in 1769 or 1770 (Richardson 1980). The chart was created when Franklin was in London as Deputy Posture General for the American colonies. One question what arose concerned why mail packets took a fortnight longer to sail to America under British captains than of merchants ships under sail by American captain Frankinsought the advice of his cousin Folger who was Nantucket ship captain then visiting London. British captains were unaware of the Gulf Stream and frequently stemmed the current. Folger sketched and Gulf Stream on a chart and Franklin had it printed. The chart continues to be an accurate Summary of the course breadth, and strength of the Gulf Stream.

Table 1.3
The Major Current

Location	*Currents East of continents warm current water*	*Currents West continents cold upwelling*
North Atlantic	Gulf Stream (including Florida Current)	Canary
South Atlantic	Brazil	Benguela
North Pacific	Kuroshio	California
South Pacific	East Australian	Peru
North Indian	Somali (during S.W. monsoon)	
South Indian	Agulhas (including Mozambique)	West Australian

The Gulf Stream carries more water than all of the Earth's terrestrial rivers combined. At Bimini in the Bahamas the flow averages about 20 milion cubic meter per second or about 1,728 cubic kilometers per day. The greatest strength of flow of the Gulf Stream occurs of the Coast of Florida. It flows in a northeasterly directs off the east coast of North America. At Cape Haltern (35° N), the Gulf Stream swings eastward and graduatly broadens as it flows towards and coast of Europe. As the Gulf Stream flows toward Europe it splits into a northern segment which flows past Norway and Spitsbergen, and southern segment which flows among the coast of Britain and France.

The Gulf Stream has an enormous impact on the marine and terrestrial environments. It acts to transport large pools of tropical marine waters. intact, complete with flora and fauna, to mid-latitude locations. In the course of its flow, the Gulf Stream undergoes meandering and large portions of the stream break off as warm-or cold-water eddies and massive "ring" (Backus et al 1981, Cresswell 1982, Wiebe 1982). Figure 1.1 schematically illustrates the distribution and movement of ring in the Atlantic. Most of the organism in the eddy gradually succumb to predation and the change of temperature as the mass of water dissipates. The contact zone between the eddies and rings and surrounding water is dynamic and a highly productive area of the marine environment which is rich in fauna and flora.

For land areas the impact of the Gulf Stream is equally profound. The Gulf Stream transport enough heat to the Scotish coast to support palm trees at Oban (56° N) and Ullapool (58° N). The flora and fauna of the coasts Scotland and Ireland have links with the Caribbean and tropical American fauna and flora. For example, sea beans are regularly found washed up on the shore. Numerous insects and a sponge from the tropics of America inhabit these coasts. Also found are a number of coastal plants, including lady's tresses, orchid, pipewort, an iris, and, Canadian blue-eyed grass.

Another organism being transported by the Atlantic marine currents is the European eel (*Anguilla anguilla*). Eels breed in the sea, starting their life cycle in the deep Sargasso Sea southwest of Bermuda and gradually moving with the Gulf Stream current toward the European freshwater rivers where they move inland to mature. They grow rapidly in the freshwater environment and upon maturity return to the sea to breed in the warm tropical water of the Atlantic. Eels and many other organisms are involved in long-distance movements that are an important part of the dynamic interplay between organisms of widely separated sites.

The depth of the sea

Water depth in the sea varies from the thickness of a film of sea water at the shoreline to more than 13 kilometers (– 6 miles) in the deep sea. Earlier in this chapter, in the analysis of the hypsometric curve, it was noted that large areas of sea floor lie between 3 and 6 kilometers (−2 to 4 miles) below sea level. These areas constitute the **Abyssal plain.**

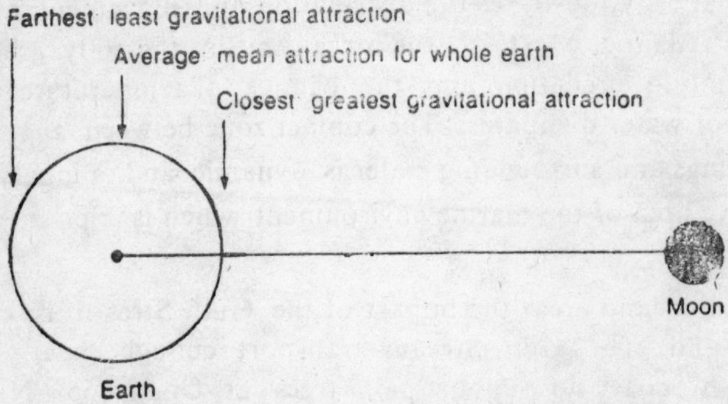

Fig. 1.2. The origin of tides in the ocean. The moon exerts a gravitational attraction on the oceans that is strongest closest to the moon causing the ocean to bulge out here. The weakest attractioh is on the side farthest from the moon, and thus the ocean bulges out here as well. The solid earth rotates beneath these tidal bulges, causing them to move in relation to the solid earth (*After F. Press and R. Siever, Earth, W.H. Freedom and Company, New York, 1982.*)

A continental shelf is nothing more than the submarine extension of a continental land mass. The **continental margin** marks the edge of the shelf ; seaward of it, the continent pinches out along the **continental slope.** Near the base of the continental slope, continental crust gives way to oceanic crust. Just seaward of this juncture is the **continental rise,** which, as will be explained more fully is prism consisting of sediment that has been transported down the continental slop. Beyond the continental rise lies the previously mentioned abyssal plain, which is the surface of a layer of sediment resting on oceanic crust. When we speak of the **deep-sea floor,** we are usually referring to the region below the continental margin ; the area above the deep-sea floor is often referred to as the **oceanic realm.**

Along the margin of the sea, **barrier islands** of sand heaped up by waves often parallel the shorline. In the protection of these elongate islands are relatively quiet lagoons or boays. **Marshes** which are formed by low-growing plants that inhabit the intertidal zone, fringe the margins of these ponded bodies of water. Here their remains accumulate as **peat**, which if buried under the proper conditions of temperature and pressure, can turn to coal. In places, the sea spreads farther inland over a continent, forming a broad, semi-isolated **epicontinental sea**. At the present time, the seas happen to stand lower in relation to continental surfaces than they have at most times during the past 603 million years. For this reason, continental seas are not well developed; Hudson Bay in erstern Canada is perhaps the best modern example. As we journey through Phanerozoic time of this book, we will examine many ancient epicontinental seas and the life that they harbored.

How is life of the ocean related to water depth ? Depth itself is probably a limiting factor for only a few species but some significant limiting factors such as light (in the case of plants) and temperature are often closely related to water depth. The upper layer of the ocean, where enough light penetrates the water to permit plants to conduct photosynthesis, is known as the **photic zone**. Although the base of this zone varies from place to place depending on water clarity, it usually lies at a depth that is greater than 100 meters (300 feet) and less than 200 meters (600 feet) below sea level it happens that 200 meters is also the approximate depth of the continental margin in most areas.

For life of the sea floor, the most profound environmental change associated with depth takes place along the margin of the ocean. At the transition between the intertidal zone and the adjacent **subtidal** sea floor, which is never exposed to the air, the biota of the intertidal zone must endure large, often rapid fluctuations in environmental conditions. At some latitudes in this zone, hot, dry conditions prevail at low tide during the summer, and yet winter chills drop temperatures below freezing. In the surf zone, where waves break along a sandy beach, the constant movement of the sand permits only a few species to survive. Species that live in this zone are exceptionally mobile and can quickly reestablish themselves in the sand in the event that they are dislodged by a wave.

Unusual conditions also characterize the deep sea, but here the environment is more stable. As we have seen, the frigid water that descends in polar regions covers the abyssal plain with waters whose temperature approaches the freezing point. In fact, cool waters characterize the entire ocean at depths below about 500 meters (1600 feet), and these waters form an environment known as the **psychrosphere**, which is inhabited by unique groups of species that are adapted to cold conditions.

In certain groups of bottom-dwelling marine organisms, different species tend to live at different depths in the ocean. The **fodaminifers**, a group of single-amoeba like creatures with internal skeletons, are one such group. They have left excellent fossil record that has provided considerable information about the water depths at which ancient sediments were deposited It is not known why different species of foraminifers tend to live at different depths; it may be that these tiny organisms are sensitive to water pressure which increases with depth.

Sea Water

Sea water is pure water in which are dissolved a variety of solids and gases. A 1000 g sample of sea water will contain about 35g of dissolved compounds, collectively called salts. In other words, 69.5 percent of sea water is pure water and 3.5 percent dissolved substances. The total amount of dissolved material is termed the ***saliniy***. Marine biologists and oceanographers, however, generally prefer to refer to salinity in terms

Table 1.4

Some Properties of Water

Property	*Compared to Other Substances*
Surface tension	The highest of all common liquids
Conduction of heat	The highest of all common liquids, except mercury

Viscosity	Relatively low viscosity for a liquid (decreases with increasing temperature)
Latent heat of vaporization; the quantity of heat gained or lost per unit mass by a substance changing from a liquid to a gas or gas to liquid phase without an increase in temperature (cal/g)	The highest of all common substances
Latent heat of fusion; the quantity of heat gained or lost per unit mass by a substance changing from a solid to a liquid or liquid to solid phase without an accompanying rise in temperature (cal/g)	The highest of all common liquids and most solids
Heat capacity; the quantity of heat required to raise the temperature of 1 g of a substance 1°C (cal/g/°C)	The highest of all common solids and liquids
Density; mass per unit volume (g/cm^3 or g/ml)	Density determind by (1) temperature, (2) salinity, (3) pressure, in that order. The temperature of maximum density for pure water is 4°C. For sea water the freezing point decreases with increasing salinity.
Dissolving ability	Dissolves more substances in greater quantities than any other common liquid

of parts per thous and, abbreviated as ‰. Thus, if a typical sea water sample has 35 g of dissolved compounds in 1000 g, it has a salinity of 35‰.

Dissolved substances include inorganic salts, organic compounds derived from living organisms, and dissolved gases. By far the greatest fraction of the dissolved material is composed of inorganic salts present as ions. Six inorganic ions comprise 99.28 percent by weight of the solid organic matter. They are chlorine, sodium, sulfur (as sulfate), magnesium, calcium, and potassium. These can be considered the major ions. An additional five minor ions add an additional 0.71 per cent by weight so that 11 ions together make up 99.99 percent by weight of the dissolved substances.

The salinity of various parts of the open ocean away from coastal areas varies within a narrow range, usually from 34 to 37 %, and average 35 %. The differences in salinity are due to differences in evaporation and precipitation. Higher values occur in tropical oceans where there is a high evaporation, and lower values in temperate oceans where there is less evaporation. In inshore areas and partially enclosed seas, the salinity is more variable and may be near 0 where large rivers discharge fresh water, to near 40 % in the Red Sea and Persian Gulf.

It has been found that whereas there are the above described variations in salinity, and hence in the total amount to dissolved salts in various areas, the ratios among the most abundant ions remain virtually constant. This is important; it means that concentrations of nearly all ions in a given sample of water can be determined by measuring only one. This fundamental relationship is the basis for the measurement of

Table 1.5

Major and Minor Constituents of 34.8 Sea Water

Ion *A. Major*	*Percent by Weight*
Chloride (Cl^-)	55.04
Sodium (Na^-)	30.61
Sulfate (SO_4^{2-})	7.68
Magnesium (Mg^{2+})	3.69
Calcium (Ca^{2-})	1.16
Potassium (K^-)	1.10
Subtatal	99.28

B. Minor	
Bicarbonate (HCO_3^-)	0.41
Bromide (Br^-)	0.19
Boric acid (H_3BO_3)	0.07
Strontium (Sr^{2-})	0.04
Subtotal	0.71
Total	99.99

salinity in sea water. Salinity can be established by measuring a single parameter, which may be the chlorinity (=chlorine concentration), for example, or the electrical conductivity or refractive index, which also depend on the salt content.

Among the remaining 0.01 percent of dissolved substances in sea water are several inorganic salts that are of crucial importance to organisms in sea water. Included here are the nutrients, the phosphates and nitrates, which are required by plants to synthesize organic material in photosynthesis, and silicon dioxide, which is required by diatoms and radiolarians to construct their skeletons. In contrast to the previous ions, nitrates and phosphates do not exist in a constant ratio with other elements or ions and tend to be in short supply in surface waters, varying in abundance as a result of biological activity. Supplies of these essential nutrients may actually become limiting to plant production in some cases.

Other substances existing as trace amounts include elements essential to life processes, such as iron, manganese, cobalt, and copper. Although present in minute amounts, they are not limiting to the existence of life. Certain organic compounds such as vitamins are also present in minute amounts, but little is known of their variation.

The salt content of sea water has a definite effect on its properities. The maximum density of pure water occurs at 4°C but with sea water, density continues to increase to the freezing point. Because of the salt content, the freezing point is also reduced from 0°C, the amount being a function of the salinity. For sea water of 35%, the freezing point is – 1.9°C. Upon freezing, however, the density decreases so that ice floats on the surface. The importance of this density increase below 4°C is that very cold and very dense surface water can be formed and sink to the bottom of the ocean basins.

Two gases dissolved in sea water are of metabolic importance : oxygen and carbon dioxide. The solubility of gases in sea water is a function of temperature ; the lower the temperature, the greater the solubility. Therefore, the colder the water, the more oxygen it can hold. Even so, the solubility of gases in water is not great. At 0°C, 35% sea water contains about 8 ml/liter O_2, whereas air has 21 ml/liter. At 20°C, 35% sea water contains only 5.4 ml/liter. The reason that deep ocean water does not become anoxic (devoid of oxygen) through

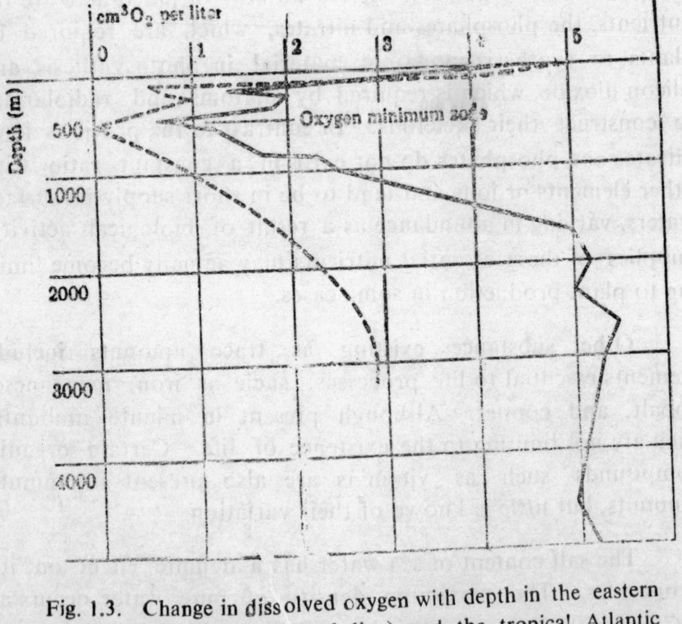

Fig. 1.3. Change in dissolved oxygen with depth in the eastern tropical Pacific Ocean (dashed line) and the tropical Atlantic Ocean (sodid), line.

biological activity is that when water sinks from the surface, it is so cold that it has a maximum amount of oxygen, more than can be consumed by the limited populations of animals in deep water. Oxygen is not distributed uniformly with depth in the ocean. A typical vertical profile of the oxygen content shows a maximum amount in the upper 10-20 m, where photosynthetic activity by plants and diffusion from the atmosphere often leads to supersaturation. With increasing depth, the oxygen content declines. This decline reaches a minimum somewhere between 500-1000 m in open ocean waters. This *oxygen minimum zone* may have oxygen values that approach zero in some areas. Below this zone, oxygen values increase somewhat with depth, but usually do not approach surface values except in the tropics. The occurrence of the oxygen minimum zone is usually attributed to biological activity depleting the oxygen, but no generally acceptable theory satisfactorily explains it.

The solubility of carbon dioxide is somewhat different from that of oxygen, since it reacts chemically in the water. Carbon dioxide is abundant in sea water, and sea water has a considerable capacity to absorb the gas. This is because carbon dioxide, upon entering sea water, reacts with the water to produce carbonic acid.

$$CO_2 + HOH \rightleftharpoons H_2CO_3$$

Carbonic acid further dissociates into a hydrogen ion and a bicarbonate ion :

$$H_2CO_3 \rightleftharpoons H^+ + HCO_3^-$$

Bicarbonate may further dissociate into another hydrogen ion and a carbonate ion :

$$HCO_3^- \rightleftharpoons H^+ + CO_3^{2-}$$

The major reservoir of carbon dioxide in the ocean is the bicarbonate ion, as can be seen from its place among the major ions in sea water. In contrast to oxygen, it is more abundant in sea water than in air. Therefore, carbon dioxide is rarely limiting to plants in sea water.

The carbon dioxide-carbonic acid-bicarbonate system is a complex chemical system that tends to stay in equilibrium. Thus, if CO_2 gas is removed from sea water, the equilibrium will be disturbed and carbonic acid and bicarbonate will shift to the left in the above equations, until more CO_2 is produced and a new equilibrium set up.

The above reactions result in the production or absorption of free hydrogen ions (H^+). The abundance of hydrogen ions in solution is a measure of *acidity*. More H^+ ions mean a more acid solution, and fewer H^- ions a more alkaline solution. *Alkaline* solutions are those that have large numbers of OH^- ions and few H^+ ions. Acidity and alkalinity are measured on a logarithmic scale of 1 to 14 units. These units are called pH units and the scale the *pH scale*. The higher the concentrations of H^+ ions on this scale, the lower the pH value. Hence, low pH values indicate acid conditions. Conversely, high pH values indicate low H^- concentrations and high OH^+ concentration. The neutral point is pH 7, where equal numbers of both ions occur.

Although pure water is neutral in pH because dis ociation of the water molecule produces equal numbers of H^+ and OH^- ions the presence of CO_2 and the strongly alkaline ions sodium, dotassium, and calcium in ocean waters tend to change this, so that sea water is slightly alkaline, usually ranging from pH 7.5 to 8.4. The carbon dioxide-carbonic acid-bicarbonate system functions as a *buffer* to keep the pH of sea water within a narrow range. It does this by absorbing H^+ ions in the water when they are in excess and producing more when they are in short supply. This is accomplished by shifting the above reactions to the right when there are too few H^+, thereby producing more bicarbonate ion and carbonate ion and to the left when there are too many, producing more undissociated carbonic acid ahd bicarbonate ion.

Basic Oceanography

This is not a book on *oceanography*, which is the discipline devoted to the study of all aspects of the physics, chemistry,

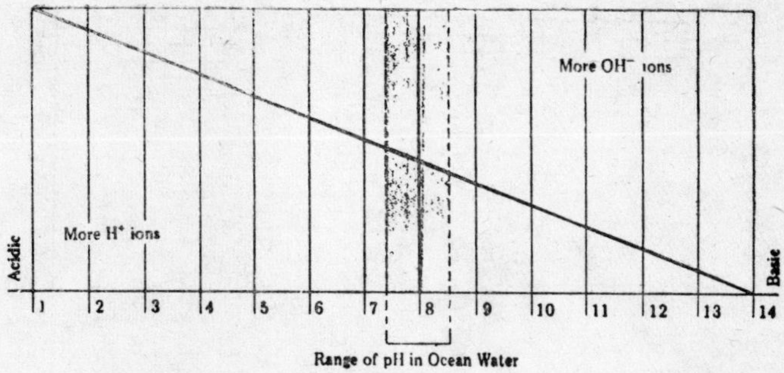

Fig. 1.4. The pH scale.

geology, and biology of the sea. We are concerned here only with giving a basic understanding of life in the sea and how it is organized and persists. However, in order to understand the ecology of various marine associations, it is necessary to know something about the structure and motion of the ocean water masses.

Geography and Geomorphology of the Oceans

Whereas all the major oceans are connected to each other, the world ocean has been separated for convenience into four major divisions, Pacific, Atlantic, Indian, and Arctic, in order of decreasing size. Projecting from, or partially cut off from, these larger oceans are smaller marginal seas such as the Mediterranean, Caribbean, Baltic, Bering, South China, and Okhotsk. The locations of the major oceans and seas. The Pacifie, Atlantic, and Indian Oceans all converge in the area around the Antarctic continent, giving a contiguous body of water around the southern continent.

The oceans are not equally distributed over the earth. Oceans cover more than 80 percent of the Southern Hemisphere but only 61 percent of the Northern Hemisphere, where most of the earth's land masses oecur.

On the margins of the major land masses, the ocean is very shallow, overlying an underwater extension of the continent called the *continental shelf*. Forming only 7-8 percent of

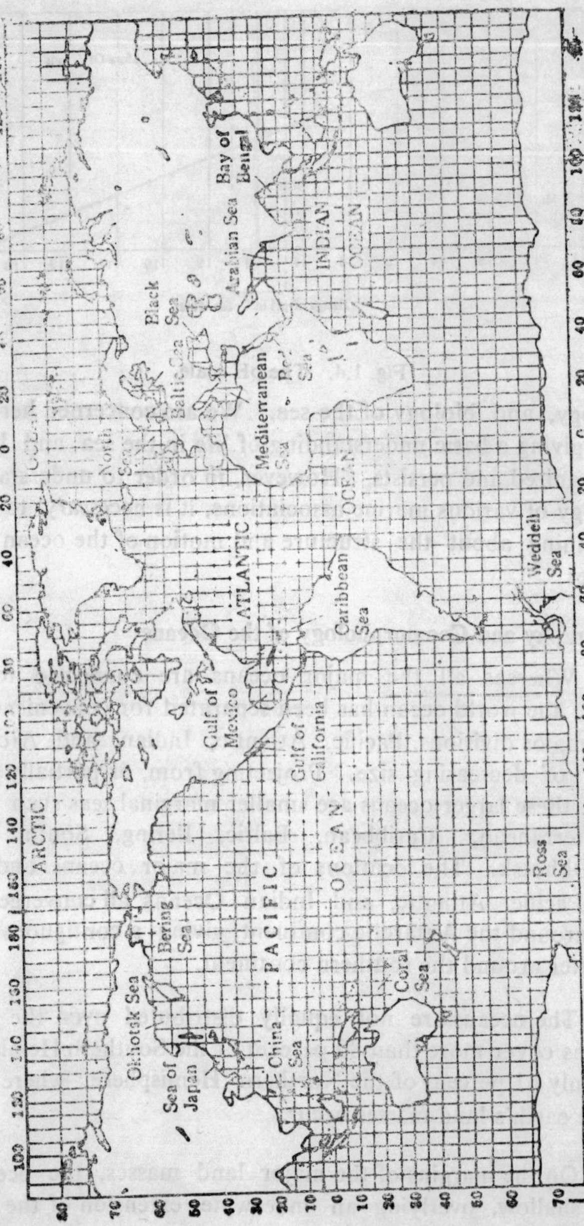

Fig. 1.5. Major oceans and seas of the world.

the total ocean area, the continental shelf slopes gently from shore to a depth of 200 m. The shelf extends offshore for up to 400 km off eastern Canada, but extends only a few kilometers offshore along most of the Pacific coast of North America. At the outer edge of the shelf, there is an abrupt steepening of the bottom to become the *continental slope.* The continental slope descends precipitously to depths of 3 to 5 km. At these depths, the bottom becomes the flat, extensive, sediment-covered *abyssal plain.* Such plains monotonously cover the floors of vast areas of the oceans at depths between 3 and 5 km.

Abyssal plains are broken in several places by various *submarine ridges.* These ridges are extensive, contiguous submarine mountain chains which have been found in all oceans. The best known is the mid-Atlantic ridge which bisects the Atlantic Ocean into east and west basins and runs from Iceland into the south Atlantic, where it links with a similar ridge in the Indian Ocean. Occasionally, the ridges break the surface to form islands. The Azores, Ascension, and Tristan da Cunha are islands formed by the mid-Atlantic ridge. These extensive ridge systems mark the boundaries of the various crustal plates of the earth and are often the sites of volcanic activity.

In certain areas, the abyssal plains are cut by deep, narrow troughs called *trenches.* Most of these trenchss lie in an are bordering the islands and continents in the Pacific Ocean. The trenches have depths from 7,000 to more than 11,000 m. The deepest area known is the 11,022 m Challenger Deep in the Marianas Trench.

Finally, there may be isolated islands and submarine *sea mounts* formed by isolated volcanic action. Such mountains, as opposed to the ridges, rise individually from the abyssal plain.

Temperature and Vertical Stratification

Temperature is a measure of the energy of molecular motion. In the world oceans, it varies horizontally with changes in latitude and also vertically with depth. Temperature is a singularly important factor in governing the life processes and the distribution of organisms. Vital life processes, collectively termed *metabolism*, function only within a relatively narrow range of temperatures, usually between 0° and 40°C. However, a few organisms are able to tolerate temperatures somewhat above and below these limits, such as the blue-green algae living in 85°C hot springs. Within the temperature range in which life processes operate, metabolism is temperature dependent. In general, for organisms which do not regulate their internal temperatures. metabolic processes are increased by a factor of two for each 10°C rise in temperature. With the exception of the marine birds and mammals, marine organisms are *poikilothermic* or *ectothermic*, meaning that their body temperatures vary with that of the surrounding water mass. *Homiothermic* or *endothermic* birds and mammals have the ability to regulate their own internal temperature, regardless of the temperature of the water mass. Most marine organisms are adapted to live and reproduce within more narrow temperature ranges than the total 10-40°C range. Since most are also poikilothermic and since the sea water temperatures vary latitudinally, the distribution of marine organisms follows closely the geographical differences in ocean temperatures.

On the basis of surface ocean temperatures and overall organism distribution, four major biogeographical zones may be established ; *palar, tropical, warm temperate, cold temperate.* Transition zones between these areas also exist and boundaries may vary somewhat with season so that these zones are not absolute.

Temperature in the oceans also has a marked variation with depth. Surface waters in the tropical regions are very warm year round, 20-30°C, and temperate zone surface waters are warm in the summer.

Below the warm surface water, the temperature begins to fall, and over a narrow depth range of 50-300 m undergoes a very rapid decline. The depth zone of most rapid temperature decline is the *thermocline.* Below the thermocline, the temperature continues to fall with depth, but at a very much slower

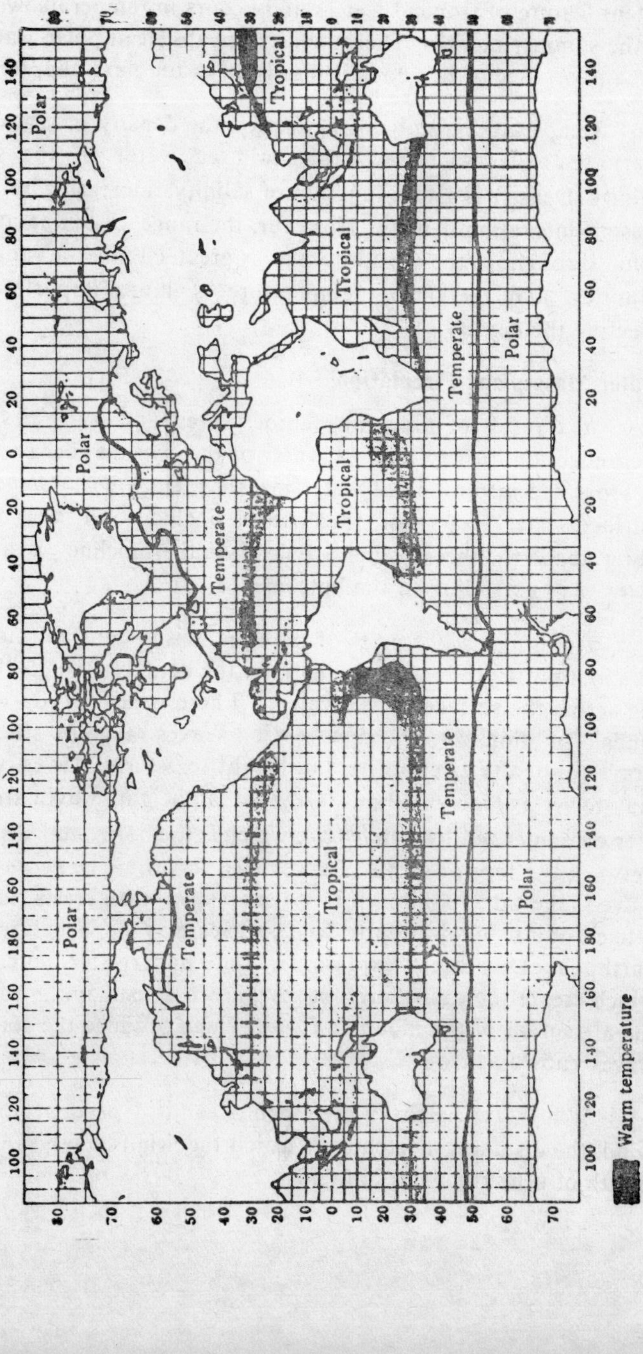

Fig. 1.6. Major biographical regions of the world's oceans, based on temperature.

rate, so that the water mass below the thermocline is nearly isothermal all the way to the bottom. The thermocline is a persistent feature of tropical waters and occurs in temperate waters in the summer months. Thermoclines are absent in polar waters. The importance of this zone is discussed in the next chapter.

Temperature also has an effect on the density of sea water. Warm sea water is less dense than cold sea water of the same salinity. Density is also a function of salinity, increasing salinity causing increasing density. However, the range of temperatures found throughout the world oceans is greater than the range of salinities, and therefore, temperature is more important in affecting the density.

Water Masses and Circulation

As a result of different temperatures and salinities and their effect on density, the sea water of the world's oceans can be separated into different water masses. The *upper water mass* (surface water mass) of the oceans includes all well-mixed water above the thermocline. Below the thermocline is a *deep water mass* extending to the bottom.

The upper water mass of the oceans is in constant motion. The motion is produced primarily by the action of winds blowing across the surface of the water. These winds produce two kinds of motion, *waves* and *currents*. Waves range in size from ripples only a few centimeters in height to storm waves which may tower as high as 30 m. Other than height, waves are further characterized by *wavelength*, which is the horizontal distance between the tops or crests of suceessive waves. The *period* of a wave is the time required for two successive wave crests to pass a fixed point. In addition to wind, waves may be generated by earthquakes, volcanic explosions, and underwater landslides, which create the destructive waves known as tsunamis, and by the attraction of the moon and sun, which produce the standing waves known as tides.

Wind wave height in the open ocean is dependent on the wind, the distance or fetch over which the wind blows, and the length of time the wind blows.

All waves behave similarly. Once generated, the waves move outward and away from the center of origin. This horizontal progression does not result in significant horizontal transport of the individual water molecules. The water molecules transcribe a circle, moving upward and to one side as a wave crest approaches and to the other side and down to near their original position as the wave crest passes. The wave form and the energy are transported horizontally. The passage of waves generates movement, not only in the surface water molecules, but also in water down to a depth approximately equal to one-half the wavelength. With each depth interval below the surface equal to one-ninth the wavelength, the orbits followed by the water particles diminish by one-half. Therefore, by the time a depth of half a wavelength is reached, the movement is almost imperceptible.

As waves enter shallow water and begin to encounter the frictional resistance of the bottom, they slow their forward motion and the wavelength decreases. As a result, they begin to increase in height and become steeper. At a point where the water depth is 1.3 times the height of the wave, it will "break," releasing the energy onto the shore.

Currents are water movements which result in the horizontal transport of water masses. The major ocean current systems are produced by a few major wind belts which succeed each other latitudinally around the world and where the winds are steady and persistent in direction. The backbone of the system is the northeast trade wind blowing from northeast to southwest between the equator and 30°N latitude and the southeast trades in similar position south of the equator, moving air from southeast to northwest. Between 30° and 60°N and S latitudes, the westerlies blow from the southwest, to the northeast in the Northern Hemisphere and to the southeast in the Southern.

These winds set the surface waters into motion, producing the slow, horizontally moving currents that are capable of transporting huge volumes of water across vast distances in the oceans. Such currents influence the distribution of marine organisms and also lead to the displacement of biogeographical

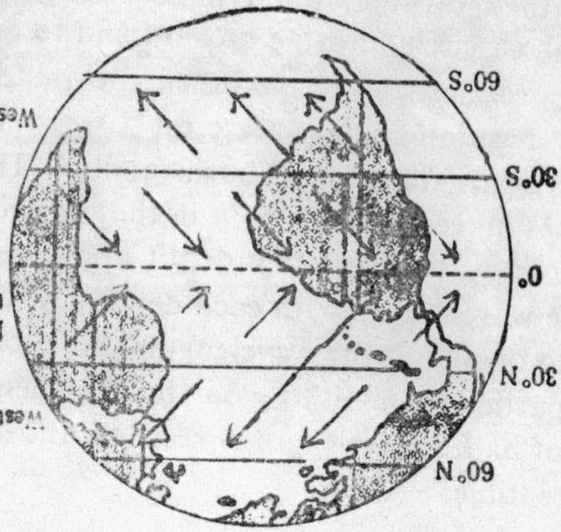

Fig. 1.7 The main wind belts of the earth and their prevailing direction of motion (arrows).

zones through the transport of worm water into colder regions and vice versa. The major currents of the world's oceans.

A comparison of Figs. 1.7 and 1.8 will show that the ocean currents do not flow parallel to wind direction. They are deflected into roughly gyres which move clockwise in the Northern Hemisphere and counterclockwise in the Southern Hemisphere. The deflections and gyres are the result of the *Coriolis force.* The Coriolis force, in turn, is the result of the rotation of the earth on its axis. The spinning of the planet imparts a deflection to moving water, displacing it to the right in the Northern Hemisphere and to the left in the Southern. Because the rotation of the earth is from west to east and because the deflection of the currents created by the trade winds imparts a water movement at the equator parallel to the equator moving from east to west, the net water movement is from east to west, pilling up water on the western side of ocean basins. As water builds up on the western sides, it meets the continental or island chain land masses and is deflected north or south as continental boundary currents. These boundary currents, in turn, moving poleward, fall under the influence of the westerly winds. The

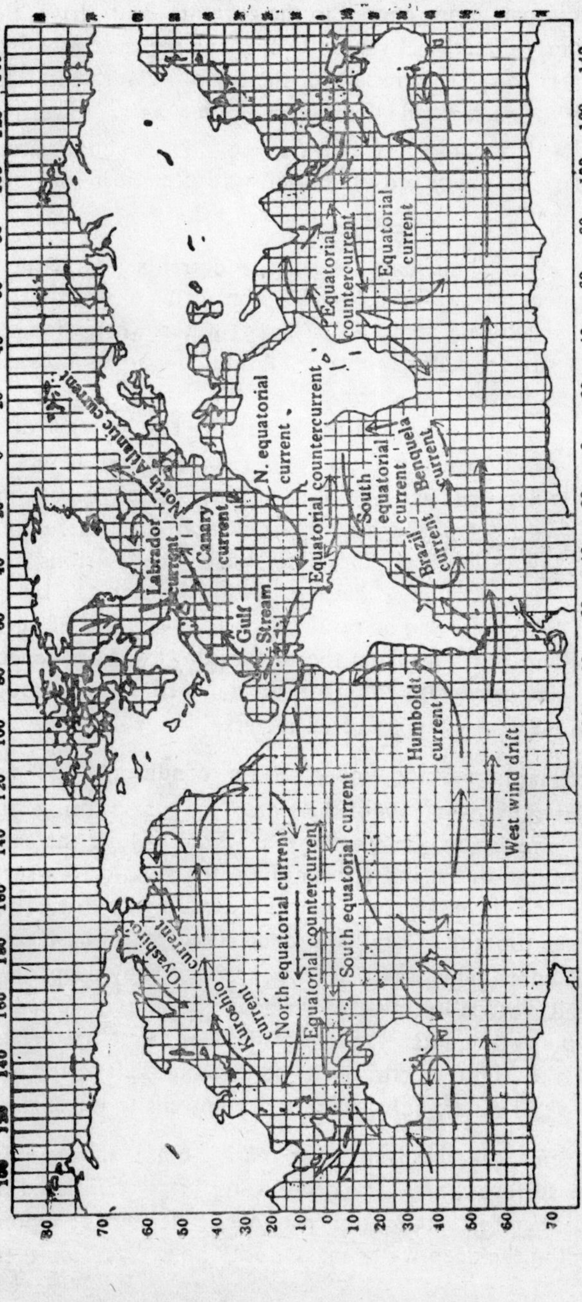

Fig. 1.8 The major ocean surface current system.

westerlies impart more energy to the currents and drive them in an easterly direction, eventually crossing the ocean basins to return water to the eastern side of the basin. Continental land masses on the eastern side deflect the moving water toward the equator, which completes the pattern. These huge circular current patterns are called *gyres* and are found in all major ocean basins.

These global currents are surface currents. How deeply do they affect the water column ? This can be estimated by noting what happens to deeper layers of water when wind stress causes the surface water to move. As noted above, because of the Coriolis force, the direction of the water movement is deflected from that of the initiating wind. The energy of the wind is passed down through the water column and it sets each successively deeper water layer into motion. Each layer, however, receives a decreasing amount of energy, and therefore, its velocity is lower. At the same time, due to the Coriolis force, each layer set in motion is deflected with respect to the one immediately above it. The result is the *Ekman spiral* of current directions and velocities from the surface downward. The depth at which wind stress fails to impart motion varies, but is roughly limited to the upper few hundred meters.

In certain areas and under certain conditions, the wind-induced lateral movements of the water may also bring about a vertical circulation or *upwilling* of water. Along the eastern margins of ocean basins, for example, the wind-driven surface currents along the continental margins flow toward the equator. At the same time, the Coriolis force tends to push these surface waters offshore. This water is then replaced by deeper water transported vertically to the surface. Similarly, along the equator, the two equatorial currents flowing west are deflected away to the right north of the equator and left to the south. To replace this water, subsurface water upwells to the surface.

Movement in the deep water mass is quite different from that in the surface mass. The deep water mass is isolated from the wind ; therefore, its motion cannot be dependent upon it. Movement in the deep water mass does, however, result from changes occurring in water at the surface. Sea water increases in density with a decrease in temperature and an increase in salinity. When sea water increases in density, it sinks. Therefore, in order to move water into the deep basin of the oceans, it is necessary to increase its density at the surface. This is accomplished in two ways. Warm water from the tropicsi s high in salinity due to evaporation. In the northern Atlantic Ocean, such warm saline water is transported out of the tropics by the Gulf Stream. In the region of Iceland and Greenland, it meets the very cold waters of the Labrador Current moving south. Mixing then cools this highly saline water, increasing its density, and it sinks to form North Atlantic deep water. Similarly, the meeting of warm water moving south with cold water in Atlantic seas also causes the sinking of water masses. A final, very high density water mass is produced in the Weddell Sea in Antarctica, where very cold water becomes more saline when winter freezing occurs. This mass sinks to become the bottom water of most ocean basins.

Since these dense waters are all cold and produced at the surface, they have large amounts of oxygen, which are then transported to the depths. Without this oxygen, deep water would be anoxic.

These messes move very slowly north and south to form the deep water of all ocenn basins. Many hundreds of years are required for these masses to move through the ocean basins. North Atlantic deep water, for example, may have been away from surface contact for hundreds of years between when it sank near Iceland and when it again surfaced in the Atarctic region. It is, however, a major contributor to the productivity of Antarctic seas, because when it surfaces, it brings with it large amounts of nutrients accumulated during the many years beneath the photosynthetic zone.

Some Ecological Principles

Because this is a text that emphasizes habitats and the ecology of marine organisms and because certain users of this book may lack a background in basic ecology, it seems necessary at this point to cover briefly some basic ecological concepts and terms.

Terms and Definitions

Ecology is the science that treats the spectrum of interrelationships existing between organisms and their environments and among groups of organisms. It is important to realize initially that living things do not exist as isolated individuals or groups of individuals. All organisms interact with others of their own species, with other species, and with the physical and chemical environments that surround them. In this interactive process, the organisms have an effect on each other and on the surrounding environment. Similarly, the various factors of the environment affect the activities of the organism. The various organisms and the environmental parameters may be further organized into various levels, each of which is somewhat broader than the preceding, A *species* is a natural group of actually or potentially interbreeding individuals reproductively isolated from other such groups. All the inviduals of a given species in an area constitute a *population*. Several species populations that tend to occur tagether in various geographical areas constitute an ecological *community*. A community or a series of communities and the surrounding physical and chemical environment together constitute an *ecosystem*. Ecosystems are the most complex entities in the above series and have many complex and interactive components. They are so large and complex that ecologists tend to study them by concentrating on their component parts, such as various communities or populations. Ecosystems, however, may be considered on a large or small scale, depending on the number of communities included and the dimensions of the surrounding nonliving environment. On the largest scale, it is possible to consider the earth as a single ecosystem comprising all the various terrestrial, fresh water and marine communities. At the other end of the scale, it is possible to consider a tidepool or fresh water pond as an ecosystem. Both have the biological and nonbiological components the definition requires.

The oceans of the world may also be considered a single ecosystem in which a series of communities is influenced by and, in turn, influences the physical and chemical factors of the surrounding sea water. This large ecosystem may, in turn, be subdivided into smaller sections or areas in which the physical and chemical parameters have differential effects on the populations of organisms, thus imposing changes in the composition and adaptations of organisms within the area subject to those effects. For example, the physical factors acting on a rocky seashore, waves and movements of the bottom, are not those that effect small organisms floating in the open ocean. Therefore, both areas contain different organisms and different adaptations. It is the aim of this book to give an understanding of the functioning of the total marine ecosystem through a consideration of the functioning of these major subsections, which form the bases of the subsequent chapters.

Ecosystem Components

An ecosystem is a functional unit of variable size composed of living and nonliving parts, which interact. The component parts and the whole system function through a sequence of operations involving energy and the transfer of energy. With few exceptions, the original energy source is the sun. Energy from the sun is captured by the *autotrophic* component, the green plants. Energy captured is stored in the chemical bonds of organic materials in the plants, which are the "food" which drives the *heterotrophic* component of the system. Hetertrophic organisms include all other life forms, which obtain their energy through consumption of the autotrophic plants or through consumption of organisms that have ingested plants. Such an arrangement of autotrophs and succeeding levels of hetertrophs is called a *trophic structure*, in which each successive consumer level is called a *tropic level.* Trophic structure is a characteristic feature of all ecosystems. The first trophic level is the autotrophic or producer level, where the energy is initially captured and stored in organic compounds. As the energy is passed from level to level in such a system, the majority of it is lost through heat and metabolic use by the organisms. The amount lost is variable, but substantial, ranging between 80—95 percent. The system thus becomes self-limiting in that, at a certain point, not enough energy remains to be passed on to sustain another level. This is visualized as an energy pyramid or trophic pyramid. In such a structure, the trophic level that consumes the plants (autotrophs, first level) includes animals called *herbivores.* Herbivores, in turn, are consumed by *carnivores*, which, in turn, are eaten by still other, often larger, carnivores. All levels above the second therefore consists of carnivores or omnivores. Within each trophic level or population, the amount of living material at any instant in time is the *standing crop*.

The final component of the trophic structure of an ecosystem is the *decomposers.* These are orgnisms, chiefly bacteria, that break down the complex organic molecules of dead organisms, releasing simple molecules usable again by the autotrophs. Decomposers act on every trophic level.

The necessary abiotic components of the trophic structure of an ecosystem are an energy source and a nutrient and water source. Plants cannot fix energy and produce complex organic molecules without sunlight for energy or without a series of inorganic nutrients, of which nitrates and phosphates are the most important.

Since ecosystems are composed of from one to many communities, each, in turn composed of many populations of producers, consumers, and decomposers, the transfer of energy through an ecosystem may follow several pathways. Each pathway that transfers energy from a given source plant or plants through a given series of consumers is callad a *food chain.* The combination of all food chains in a given community or ecosystem is called a *food web.* The food web is thus a summary of all the pathways by which energy moves from one level to another through a community or ecosystem.

Biogeochemical Cycle

Among many of the chemical elements and compounds

in ecosystems there is a cycling back and forth between organisms and the physical environment. Such repeated transfer are called *biogeochemical cycles.* A few of these cycles involve chemical compounds vital to the continued maintenance of life in the ecosystem and are therefore extremely important. Perhaps the most significant are the cycles involving carbon, nitrogen, and phosphorus.

In all these cycles, there is a major reservoir or pool of the element from which the element is continually moving in and out as it passes through organisms. Each cycle also contain a sink into which a certain amount of the chemical passes and from which it is not recycled in the normal course of events. Over long periods of time, the loss to the sink may become limiting, unless the sink can be tapped again. This later situation usually develops through geological action that acts to move the sink into an area where organisms, erosion, or other factors release the elements. Biogeochemical cycle tend, finally, to have self-regulating feedback mechanisms that keep the cycles in equilibrium.

In the carbon cycle, the reservoir is in the form of CO_2 which exists in water in the carbonic acid-bicarbonate-carbonate system. It is fixed into organic compounds by plants, transferred to animals through herbivory and predation, and returned to the reservoir via respiration and bacterial action. A non-renewable loss occurs when heavy carbonate materials such as shells are deposited in deep oceans as calcareous sediments.

In the phosphorus cycle, the major reservoir is in phosphate rock. Here, erosion bringes the chemical into the water where it cycles through animals and plants and is returned to the general circulation through decay and excretion. The loss is to deep sediments, which are unavailable to organisms.

Air is the major reservoir in the nitrogen cycle. Nitrogen gas is not used by most organisms and must be converted into a nitrogen compound to enter the cycle. This is normally done by bacteria and/or algae. Useful nitrogen compounds may also be produced through volcanic activity. Once in compounds, the nitrogen cycle through organisms with some loss to deep sediment.

Biotic Structure of Ecosystems

Communities and ecosystems have the same trophic levels throughout the world. However, the species constituting each differ among geographical areas and, in some area, each level may have more or fewer species than another. This implies that ecologically equivalent species replace each other geographically and that the species structure of communities is variable, even of the trophic structure is not. For example, sea grass beds in the tropics may have several species of grasses, whereas those in the temperate zone but one. They are different species, but ecologically fulfill the same role, as autotrophs. Species that are ecological equivalents are those that perform the same function or role in a given community. They are therefore said to occupy the same *ecological niche.* Ecological niches may be board or narrow. Narrow niches mean that the role or function in the community or ecosystem has been more finely subdivided and the species is therefore more specialized. Broad niches, on the other hand, mean the function is more generalized. Species may therefore be either *specialists* or *generalists* with reference to their niches. In our precedtng example, the autotroph niche is broad in the temperate zone and therefore occupied by a single generalist grass ; in the tropics, several more specialized grasses subdivide the niche.

In contrast to niche, an ecological *habital* refers to the place where an organism is found. Therefore the small copepod crustacean *Calanus finmarchicus* has as a habital the surface waters of the North Atlantic Ocean, but its niches is that of a herbivore feeding on the small plants in the area.

Most communities have a characteristic species structure that consists of a few species that are abundent and a larger number of species that are rare. The abundant species are usually called *dominates* and are often used to characterize a community. Thus, we can speak of a "mussel bed" on the rocky shore reflecting and dominance by the mussels *Mytilus edulis* or *Mytilus californianus*. This pattern of few common and many rare species is true at each trophic level. It is also true whether the community has many or few species in total. Species structure in ecological communities may be measured in several ways. *Species richness* is a simple listing of the total number of species in a community or trophic level. *Spacies diversity* is a measure that combines into a single figure both the number of species (richness) and the distribution of the total number of individuals among the species (evenness). It is expressed through various mathematical diversity indices which need not concern us here. High species diversity is generally taken to be indicative of benign, stable environments and low diversity of stressful, fluctuating environments.

Each species in a community has certain tolerances with respect to each and every environmental factor. If the limit of tolerance is exceeded in some area by a given factor, say temperature, the species will be absent. Similarly, each species requires a certain minimum amount of various materials. If the concentrations of these necessities, such as nitrate, fall below the minimum, the species disappears. More importantly, if any *single factor* exceeds the tolerance level or any single necessary substance is reduced below the minimum, the species will be eliminated. This is true even if all the other factors and substances are favorable. This is known as *the law of the minimum*.

Since these limiting factors vary for each species and since communities are defined on the basis of recurrent groups of species, it follows that the boundaries between communities, representing changes in various environmental factors, are also not sharp. The result is areas of transition between adjacent communities where species gradually drop out and others come in. Such boundary, zones are called *ecotones*.

Communities are not static units. They change in structure and composition with season and over longer periods of time. Terrestrial communities tend to change in an orderly fashion over periods of many years until they reach a stage which perpetuates itself indefinitely as long as the climate does not change or there is no disturbance. In such a sequence, each community modifies the environment, in turn making it suitable for the next community. This orderly process of community change controlled through modification of the physical environment is called ***ecological succession*** and the terminal, persistent community the *climax.*

Ecological Control and Regulation

Populations, communities, and ecosystems are all regulated by various factors. The major controlling factors for ecosystems and communities are energy, the physical factors collectively termed climate or environment, and the interaction among various species that compose the systems. Ultimately, nearly all systems on earth are limited by the amount of energy available from the sun. However, tolerance limits of various species to abiotic factors such as temperature, light, nutrients, and salinity also limit the extent of populations and communities in the oceans. The final aspects of regulation and control considered here are those dealing with the interaction among populations that act to keep populations within limits.

Biologists have long known that all species possess the reproductive potential to produce much longer populations than are observed under natural conditions, and that if population explosion occur, they are quickly reduced. What are the biological factor that exert control ? They are grouped under predation, campetition, and disease and parasitism.

Competition is an ecological term referring to the interaction among organisms for a necessary resource that exists in short supply. Competition may be ***intra-specific*** (among individuals of the same species) or ***inter-specific*** (among individuals of different species). Competition may be for any number of things, but usually is limited to items such as light, food, nutrients, water, and space. In a competitive interaction, either the competitors manage to share the limited resource or one excludes the other. In the first case, both individual are hampered, which inhibits their growth, development, and reproduction, thus limiting the numbers. In the second case, one individuals is eliminated, again controlling population. Interspecific competition is usually between two closely related species and has led to the *competitive exclusion principle,* which say no two species with exactly similar requirements can coexist in the same place at the same time ; that is, complete competitors cannot coexist. As numbers increase, competition usually increases, because the limited resource becomes scarcer. This increased competition increase the stress on animals and plants and absorbs energy otherwise used for reproduction, thus limiting populations. Direct competitive interactions among organism are rarely observed and are usually inferred from changes observed in population numbers and distributions in nature.

Predation can be defined as the consumption of one species by another. The animal that is the consumer is the *predator* ; the victim the *prey.* A special case concerns those animals that consume plants. They are called *herbivores.* A *grazer* is an animal which feeds on plants or sessile animals. Predators and herbivores vary considerably in their ability to regulate the numbers of organisms they consume. In some cases, the predator may be the most important factor in regulating numbers of a prey species. In other cases, a predator may have little effect on a prey population. In the first case, the removal of the predator will have a marked effect on the prey population, causing it to increase dramatically. In the latter case, predator removal has little effect on the prey population. We shall see examples of both these situations in the following chapters. A special ease of predation concerns those predators that have a profound effect, not only on their prey population, but also on the entire community of which they are a part. In these cases, the removal or depletion of the numbers of the predator has the effect of causing great changes in the presence and abundance of many species in the community, most of which are not the prey of the predator. As a result, the entire community structure may be charged. Such predators are called *key industry* or *keystone* species. An example of such a species is the starfish *Pisaster ochraceus.*

Parasitism and disease are the final biological controls on populations. As with predation, they may exert strong control or they may have little effect. *Parasites* are organisms living in or upon other organisms from which they derive nourishment and shelter. Many, perhaps all, marine organism have parasites, but we known considerable less about the role of parasites, and disease in regulating populations of marine organisms than we do for terrestrial organisms. Perhaps the best documented change induced in a marine community by a disease is the loss of the eelgrass beds in the Atlantic Ocean in the 1930s.

Comparison of Terrestrial and Marine Ecosystem

It is quite true that ecological principles apply equally whether one is dealing with terrestrial, fresh water, or marine ecosystems. However, the special conditions prevailing in salt water that are the result of the physical and chemical properties of water, have, directly or indirectly, channelled evolution and adaptation of marine organisms. The result has been some striking differences in the organization of marine communities when compared with terrestrial communities. These will be considered next.

Physical and Chemical Differences

Sea water has several physical features that have a profound effect on the organization of marine communities. They are its greater density with respect to air and its ability to absorb light. The greater density of sea water means that relatively large organism and particles can float around in it. This is not possible in air. One significant result of this is that the marine ecosystem has evolved a whole community of small organism that are perpetually afloat, the *plankton.* No such comparable community exists terrestrially floating in the air. As a result, all communities in the sea are bathed in a medium that itself contains a community. The presence of this suspended community has, in turn, led to the evolution of animals that are adapted to filter organisms and particles out of sea water. These *filter feeders* are unique to aquatic systems. There is nothing really comparable on land. The closest analog would be spiders, which capture flying insects in their webs. Futhermore, since the waters of the oceans are in perpetual motion, many of these filter feeders are *sessile* (fixed in place), extending their filtering nets up into the moving water mass. Sessile animals are not possible in the terrestrial environment because the bathing medium flowing by contains no suspended food organisms as particles. Finally, marine organisms, especially the sessile species, have envolved motile larval forms. Such motile. larvae, when put into the plankton, permit dispersal of the species.

As a result of the suspension of a whole community in the bathing medium, it is more difficult to have isolated communities and isolated species distributions in the sea.

In contrast to air, water strongly absorbs light. As a result light entering water can only penetrate to certain depth before it is complete absorbed. Although this depth varies, it is sufficiently shallow (100 m) that the great majority of the volume of water in the oceans is without light. This means that plant life and primary productivity are limited to an extremely narrow band near the surface. Whereas terrestrial communities, with the exception of caves, all possess sufficient light for plant growth, marine communities exist without the benefit of an autotrophic component.

Not only does water absorb light, it does so differentially, depending upon the particular wavelength. The result is that certain wavelengths penetrate dee per than others. This is not true on land, where the same spectral composition of light impinges at all levels in any community.

Another physical factor indirectly responsible for differences between marine and terrestrial organisms is *gravity*. Because marine plants and animals are buoyed up by water, they do not need to have a significant amount of their biomass invested in structural material such as skeletons or cellulose in order to hold themselves erect against the force of gravity. Similarly, where movement is concerned, terrestrial animals must raise their mass against the force of gravity for each step. Such movement requires significantly more energy than swimming movements do in aquatic organisms and hence, more energy storage.

This absence of large amounts of structural material and energy storage in the bodies of marine organisms is reflected in the differences in the dominant biochemical compounds found in terrestrial and marine organisms. Among terrestrial organisms, the dominant compounds are carbohydrates; among marine organisms, the predominant material is protein. Living organisms that are composed primarily of carbohydrate are long lived, slow growing, and rich in stored energy. Protein-dominated organisms, on the other hand, are rapid growing and without significant energy storage.

A final physical difference between terrestrial and marine systems concerns oxygen. In air, oxygen constitutes a nearly constant 21 percent of the volume throughout the world. However, water holds less oxygen, and furthermore, its concentration can vary with temperature and salinity. Thus, marine communities are subject to variations in oxygen not found among their terrestrial counterparts.

Structural and Functional Differences

One striking difference between terrestrial and marine communities is the insignificance in the latter of large macroscopic plants. Terrestrial communities are universally dominated by large flowering plants that are persistent and long lived. With the exception of certain large kelp plants, the communities of the sea have no large plants and the dominant autotrophs are microscopic plants of various groups of algae. In turn, this means that the dominant herbivores of the sea are also small, often microscopic, animals, in contrast to the largebobied herbivores common to terrestrial communities. There are, for example, few herbivores in the sea to compare in size with the antelopes of the African plains or the bison of the North American grasslands. The dominant marine herbivores are microscopic crustaceans called copepods.

Another characteristic of terrestrial plants is that significant portions are composed of rigid structual materials such as wood and fiber, which themselves are relatively undigestible by most herbivores. As a result, herbivore grazing in terrestrial communities rarely removes significant amounts of the plant community matrix. By contrast, the small herbivores of the seas usually consume the entire plant and a major difference between terrestrial and marine ecosystems is the ability of the herbivores to completely remove the plants.

Terrestrial communities are characterized by a matrix of long-lived plants and a fauna that is generally shorter lived. In the seas, at least among those communities associated with the sea bottom, the plants, if present, are short lived, and the matrix of the community is of relatively long-lived animals, Thus, one speaks of redwood forests, oak forests, tail grass prairies, on land, but of coral reefs, clam beds, oyster reefs and mussel beds in the seas.

As a result of the microsopic nature of most marine plants and herbivores, the majority of largr animals in the sea carnivores. Large marine animals are therefore on higher trophic levels than large land animals. From the standpoint of human beings, harvestable food in the form of animal protein is available primarily from large herbivores on land (second trophic level), such as cattle and sheep, whereas harvestable food from the sea is usually in the form of carnivores from the third or forth trophic levels, such as salmon, tuna. and halibut. A corollary of this is that most marine food chains have about five links (steps) to reach the top carnivore, whereas terrestrial food chains tend to be shorter, averaging three links to reach the top carnivore.

There is also a question about the efficiency of energy production and transfer between terrestrial and marine ecosystems. Generally, the production of organic material is higher in terrestrial ecosystems than in marine, but the efficiency of transfer from first to second trophic levels is higher in marine ecosystems. Thus, although starting out with less energy fixed marine systems appear to lose less with the first transfer. Attempts to further assess the efficiency of energy transfer in marine food chains are complicated because of the various feedback mechanisms which tend to "blur" the trophic level to which to assign the animal. The primary group responsible for this blurring of the trophic level is the filter feeders. These animals tend to feed upon suspended particles, and the distinction as to what is retained on their filters is made on the basis of size, not whether the particle is animal or plant or living or dead. This means they often take in both plant and animal material and feed on several trophic levels. A second reason for blurred trophic levels is the tendency of marine species to switch positions in the trophic hierarchy. Thus, juvenile fishes, for example, may feed on herbivorous copepods, but when adult change to feed upon other fishes higher in the trophic spectrum. An additional problem is the ability of various marine organisms to switch trophic levels as adults. Thus, for example, salmon may feed on squid at one time, but switch to lantern fish at another, thus shortening the food chain since lantern fishes are themselves the food of squid. Similarly, euphausiid crustaceans ("krill") may be herbivores in shallow water and carnivores in the open ocean.

The above-mentioned differences between terrestrial and marine ecosystems should be kept in mind as we explore the various habitats and communities of the sea in the remainder of this book.

Division of the Marine Environment

The marine ecosystem is the largest aquatic system on the planet. Its size and complexity make it difficult to deal with as a whole. As a result, it is convenient to divide it into more manageable subdivisions, each of which can then be discussed in terms of the ecological principles that govern the adaptations of the organisms and the organization of the communities. No universally acceptable scheme of subdivision of the marine environment has yet been proposed. The one followed here is modified from Hedgpeth (1957) and has enjoyed widespread use among biologists for 20 years. It therefore probably comes closer than any to a mutually acceptable division.

Major Subdivisions of the World Ocean

Beginning with the waters of the open ocean, there are subdivisions that can be made in both the vertical and horizontal directions. The entire area of the open water is termed the pelagic realm; pelagic organisms are those that live in the open sea away from the bottom. This is in contrast to the benthic realm, which is a general term referring to organisms and zones of the sea bottom. Honzontally, the pelagic realm can be divided into two zones. The *neritic* zone encompasses the water mass that overlies the continental shelves. The oceanic zone includes all other open waters. Progressing vertically, the pelagic realm can be further subdivided. Two schemes are possible. The first is based upon light penetration. The *photic* zone is that part of the pelagic realm which is lighted. Its lower boundary is the limit of light penetration and varies in depth with clarity of the water. Generally, the lower boundary is between 100 – 150 m. A synonym for this zone is the *epipelagic* zone. Because it is the zone of primary production in the ocean. The permanently dark water mass below the photic zone is termed, the *aphotic* zone.

The pelagic part of the aphotic zone can itself be subdivided into zones that succeed each other vertically. The 'mesopelagic, is the uppermost of the aphotic areas. Its lower boundary in the tropics is taken to be the 10°C isotherm, which may be at 700 to 1000 m, depending on the area. Next is the bathypelagic, lyingg between 10°C and 4°C, or in depth between 700 and 1000 m and 2000 – 4000 m. Overlying the plains of the major ocean basins is the abyssal pelagic, which has its lower boundary, at about 6000 m. The open water of the deep oceanic trenches between 6,000 – 10,000 m is called the hadalpelagic.

Corresponding to the last three pelagic zones are three bottom or benthic zones. The bathyal zone is that area of bottom encompassing the continental slope and down to about 4000 m. The abyssal zone includes the broad abyssal plains of the ocean basins between 4000-6000 m. The hadal is the benthic zone of the trenches between 6,000-10,000 m. This entire aphotic area, although encompassing by far the largest volume and area of the world's oceans, is little known.

The benthic zone underlying the neritic pelagic zone on the continental shelf is termed the sublittoral or shelf zone. It is illuminated and is generally populated with an abundance of organisms constituting several different communities, including sea grass beds, kelp forests, and coral reefs. The majority of this area is covered in but the specialized coral reef communities are discussed.

Two transitional areas exist, one between the marine environment and the terrestrial and the other between marine and fresh water. The intertidal zone or littoral zone is that shore area lying between the extremes of high and low tide; it represents the transitional area from marine to terrestrial conditions. It is a zone of abundant life and is well studied. It is well to remember that the above classification scheme is not absolute, but is used for convenience, and that organisms and communities may extend into two or more zones. This is particularly true for those organisms that are powerful swimmers and divers.

2

Origin of Earth

Life Began on Earth. There were no biological Big Bangs, nor extraterrestrial cultivators, just evolution from plausible nonliving beginnings." That is how John Scott (1981, 153), a biochemist at the Manchester Medical Schoo' in England, stated the basic assumptions scientists generally make about the origin of terrestrial life. The question is, how did evolution from nonliving beginnings proceed ? This chapter will attempt to answer this question, following theoretical developments that have gained widespread support from the scientific community since the late 1970s.

The story begins with the physical and chemical events that are believed to have taken place on the surface of the young Earth roughly 4 to 4½ billion years ago. At that time violent and nearly incessant volcanic eruptions occurred at many places, while at the same time the final accretion phase of the planet's formation drew to a close and its surface layers began to reach some semblance of equilibrium. The eruptions spewed forth large quantities of water, carbon dioxide, molecular nitrogen, and many other molecules, from which the first atmosphere and the juvenile ocean formed. Driven by energy from sunlight, lightning, volcanic heat, and meteorite impacts, the inorganic molecules reacted chemically with each other and produced a great variety of organic molecules—amino acids, sugars, lipids, the bases of nucleic acids, and many more,

Gradually these molecules accumulated in the waters of the Earth until, in the words of John Haldane, "the primitive oceans reached the consistency of hot dilute soup" (Dickerson, 1978, 31). Today scientists believe that the temperature of the primitive ocean was probably close to the freezing point of water. Furthermore, its content of organic molecules may not have been as concentrated as Haldane had envisioned it, at least not throughout most of its volume. Nevertheless, many people still refer to it as the "primordial soup."

At present we are far from understanding all of the chemical reactions that took place on the young Earth. It seems that simple organic molecules assembled into large and more complex molecules similar to those found in today's organisms, as suggested by laboratory experiments. For instance, amino acids probably assembled into peptide chains, and nucleotides assembled into short strands of RNA and other kinds of nucleic acids. Surfaces of clays, lava, rocks, sand, and other readily available substances may have served as catalysts facilitating these assembly reactions.

According to the theory of the origin of life that I am presenting here, short strands of RNA were the first molecules in the primordial soup that carried information, albeit very little, and they are regarded as the starting point of the evolution toward cellular life. (For alternative theories favored by some biochemists, see Cairns Smith 1985 and Dyson 1985.) The short strands of RNA were capable of self-replication. In the process mistakes were made, so that the copied RNAs frequently differed from the original ones with regard to nucleotide sequence and length (recall that RNA nucleotides are of four types, with the bases A, G, C and U).

Some of the RNA molecules were more successful than others in surviving and replicating themselves under the prevailing conditions. Thus, the two components that form the basis of the Darwinian theory of evolution—random creation of variation and natural selection – may have been introduced very early among the chemical reactions in the primordial soup. As chemical evolution continued, different sets of RNA molecules coupled together into cooperative units. Some of the RNAs carried instructions for the assembly of primitive enzymes (peptide chains), while others acted as catalysts and contributed to the actual assembly of enzymes. Enzymes, in turn, helped in the replication of RNAs. Eventually, some of the coupled units of RNAs and their enzymes became enclosed by membranes and the first primitive cells—the protocells – were born. Life had emerged from among the random and spontaneous chemical reactions in the primordial soup.

This brief summary outlines the key components of the events that many biologists and chemists believe may have been central to the origin of life on Earth, and the remainder of this chapter will fill in some of the details. At present this theory is based on many assumptions and contains many gaps. It is not based on any direct evidence dating back to the primordial soup, for none has survived.

Furthermore, laboratory experiments that attempt to simulate early Earth conditions and to reproduce the chemical reactions that took place then have been only partially successful in telling how the protocells arose. Scientists think they understand in general terms how the Earth's atmosphere and ocean came into existence. They believe they know something about the formation of organic monomers from inorganic raw materials and the assembly of polymers from monomers. However, they still have only a very limited understanding of the emergence of order and information among the chemical reactions in the primordial soup. And they know virtually nothing about the evolution from those initial chemical reactions to the formation of the first cells. The Nobel prize-winning German biochemist Manfred Eigen (1981, 114) characterized very aptly our current state of knowledge of how life began: "Anyone attempting [to re-create life] would be seriously underestimating the complexity of prebiotic molecular evolution. Investigators know only how to play simple melodies on one or two instruments out of the huge orchestra that plays the symphony of evolution."

We should not be discouraged by this lack of knowledge.

Let us accept the problem of the origin of life as one of the great challenges facing science today. Let us also accept the fact that time inevitably diminishes and sometimes erases evidence of long-ago events. Hence, our first task is to discover the fragments.

Table 2.1

Partial Listing of Molecnles

CH_4	methane	H_2O	water
CO	carbon monoxide	H_2S	hydrogen sulfide
CO_2	carbon dioxide	NH_3	ammonia
$CO_3^{=}$	carbonate	NO_3^-	nitrate
C_2H_4	ethylene	N_2	molecular nitrogen
C_2H_6	ethane	O_2	molecular oxygen
HCN	hydrogen cyanide	$PO_4^{\equiv}$	phosphate
H_2	molecular hydrogen	$SO_4^{=}$	sulfate
H_2CO	formaldehyde		

Origin of the Earth's Atmosphere and Ocean

The earthquakes, volcanic outbursts, and bombardments by meteorities, which ravaged the young Earth so regularly, did not just churn the crust and produce large lava flows. As hot lava reached the surface and meteorites heated their impact areas to incandescence, gases that had been trapped in the rocks burst into the open in huge amounts—gases of water (H_2O), carbon dioxide (CO_2), molecular nitrogen (N_2) and, in lesser amounts, of molecular hydrogen (H_2), carbon monoxide (CO), methane (CH_4), ammonia (NH_3), hydrogen sulfide (H_2S), and many others, Our planet was acquiring its first atmosphere,

This kind of outgassing (releasing of gases) can still be observed today, although at a considerably diminished rate, in the hot springs and geysers of Yellowstone National Park, active volcanoes such as Mount St. Helens, and many other places of geothermal activity. For example. volcanic eruptions are usually accompanied by the emission of thick and often foul-smelling clouds of gases that billow for miles into the atmosphere. The gases originate from within the lava. They escape into the open when the hot lava reaches the surface of the Earth and is no longer subjected to the high pressures deep below the ground. Quite often the gases bubble forth in ways that give the resulting rocks a frothy appearance and make them lighter than water. During some eruptions the gases burst forth so violently that they shatter the lava into fine-grained dust known as ash, which may be thrown hundreds or even thousands of kilometers into the surrounding areas.

Table 2.2

Relative Abundances (in Percent) by Mass of the Most Common Elements in the Sun, Entire Earth's Earth's Crust, and Human Body

Element	*Sun*[1]	*Entire Earth*[2]	*Earth's Crust*[2]	*Human body*[3]	
Hydrogen	73.6	—	—	10	(63)
Helium	24.8	—	—	—	—
Carbon	0.29	—	—	18	(9.5)
Nitrogen	0.093	—	—	3.1	(1.4)
Oxygen	0.77	30	46	65	(25)
Neon	0.12	—	—	—	—
Sodium	0.0029	—	2.1	0.1	(0.03)
Magnesium	0.046	13	4	0.04	(0.01)
Aluminium	0.0049	1.1	8	—	—
Silicon	0.069	15	28	—	—

(Contd)

Phosphorus	0.0007	—	—	1	(0.2)
Sulfur	0.038	1.9	—	0.3	(0.05)
Chlorine	0.0011	—	—	0.2	(0.03)
Argon	0.018	—	—	—	—
Potassium	0.0003	—	2.3	0.4	(0.06)
Calcium	0.0057	1.1	2.4	2	(0.3)
Iron	0.16	3.5	6	—	—
Nickel	0.0084	2.4	—	—	—

Note : The last column, given in parentheses. refers to abundance by number relative to a total of 100.

Sources : [1]Noses 1982; [2]Press and Siever 1978, 14; [3]Keeton 1976, 28.

Water

Despite the heat that accompanied outgassing on the early Earth, the average temperature of the atmosphere was probably not far above the freezing point of water, for the energy output of the Sun—our planet's major source of heat—was then only about two-thirds of what it is today. Hence, the water, which was the most abundant molecular species emitted by volcanism and meteorite impacts, did not remain in gaseous form for long. It condensed into rain or snow and fell to Earth. Very likely the early rains came down in torrents , accompanied by storms, lightning, and thunder. The first rivers began to flow, glaciers and ice caps accumulated in the polar regions, and low-lying basins filled with liquid water to become our planet's juvenline ocean.

Carbon Dioxide

Volcanic outgassings on the early Earth poured forth not only water but also many other gases. The most abundant gas after water was carbon dioxide. This presented a potential danger for the eventual origin and evolution of life on Earth be ause a planetary atmosphere that contains substantial amounts of CO_2 heats up to high temperatures by the *greenhouse effect*. This effect comes about as follows. Solar radiation, whose wavelengths correspond mainly to the colors blue through red, penetrates relatively easily through the atmosphere and is absorbed by the ground. The absorbed energy heats the ground, which reradiates it in the infrared part of the spectrum. Infrared (IR) radiation has long wavelengths and, unlike the original solar radiation, much of it becomes absorbed by the CO_2 in the atmosphere. The CO_2 molecules reradiate this absorbed radiation, with some of it being sent back toward the ground. The net result is that the IR radiation does not readily escape into space but becomes trapped in the atmosphere, heating it as well as the ground.

Fortunately, the greenhouse effect had only a small effect on the young Earth's surface temperature because the rains washed the CO_2 out of the atmosphere before it could accumulate to dangerous levels. This happened because CO_2 dissolves in liquid water (but not in gaseous water) and forms carbonic acid (H_2CO_3) :

$$CO_2 + H_2O \text{ (liquid)} \rightarrow H_2CO_3$$

The rains and rivers brought the carbonic acid in contact with the ground, where it leached positive ions of calcium (Ca^{++}) and magnesium (Mg^{++}) out of rocks and combined with them to form limestone ($CaCO_3$) and dolomite [$CaMg(CO_3)_2$]. The limestone and dolomite then became deposited as sediments on the ocean floor. Thus, most of the original CO_2 that had been released into the atmosphere became locked up in sedimentary rocks. (Note that after shale and sandstone, limestone constitutes the most abundant sedimentary rock type of the present-day terrestrial crust.)

The third most abundant gas released by volcanism and meteorite impacts on the early Earth was molecular nitrogen, N_2. It is chemically rather inert and did not condensed into liquid form at the temperatures that prevailed on Earth, nor did it combine with water. It stayed in the atmosphere and, in the course of time, accumulated to become its dominant molecular constituent.

Oxygen

Interestingly, molecular oxygen (O_2)—which today makes

up 21% (by number) of our atmosphere and is essential for the survival of all life (except some bacteria) – was present in only very low concentration in our planet's original atmosphere. Had it been abundant, it would have rusted the iron and other metals. The oldest known rocks show no consistent evidence of such rusting. Only rocks younger than about 2 billion years do and, thereby, attest to the abundant arrival of molecular oxygen in the atmosphere by that time. Molecular oxygen was released in appreciable quantities into the atmosphere only after some organism acquired the ability to split water molecules as part of their photosynthetic activities.

The absence of O_2 from the early atmosphere was a fortuitous circumstance. Had it been present in large amounts, its high chemical reactivity would have made the development of life very difficult, if not impossible. Even today O_2 is inherently poison to all organisms. They can withstand its reactivity only because they possess special molecules that chemically bind with O_2 and let it react in slow, controlled steps.

Without O_2, the atmosphere of the young Earth did not have the ozone (O_3) layer that it has today. This layer is crucial for our survival, for it filters out most of the ultraviolet (UV) radiation from the Sun and prevents it from causing harmful chemical reactions in our cells, such as burns and damage to DNA and RNA (which may lead to cell death or cancer). In the absence of an ozone layer on the young Earth, UV radiation penetrated freely to the ground. But instead of being harmful, it provided the energy for triggering numerous chemical reactions that were important for prebiotic evolution, as discussed in the following sections.

The Next Step

Very likely, the first outgassings on the primitive Earth began long before the planet was fully formed and they continued with great intensity for hundreds of millions of years thereafter. Only when the initial burst of earthquake activity, volcanism, and meteorite impacts had abated. appr oximately

4.0 to 3.5 billion years ago, did the outgassing slow down. Geologic evidence indicates that by then the oceans covered large areas of our planet, and we may assume that the atmosphere had acquired a substantial fraction of its present content of N_2. Most of the outgassed CO_2 had been washed out of the atmosphere and locked up in sedimentary rocks, but not all of it. Some of the CO_2 together with other gases that had been discharged in small amounts, such as H_2, CH_4, NH_3, and H_2S, embarked upon a rather different and much more interesting evolutionary course. They reacted chemically with each other to form molecular structures of ever-increasing complexity.

Organic Chemistry on the Early Earth, Phase One : Synthesis of Monomers

There are good reasons to believe that conditions on the surface of the young Earth were quite suitable for the occurrence of a great variety of chemical reactions and the production of many kinds of molecules. Volcanic outgassing supplied ample amounts of atomic and molecular raw materials. The temperature was low, but not so low as to freeze water everywhere. Lightning, UV radiation from the Sun, geothermal heat from volcanoes and hot springs, and meteorite supplied plenty of energy for driving the reactions. At the same time, the ocean and ground offered protection from too much UV radiation and heat that might have destroyed the molecular products.

Among the molecular raw materials on the surface of the young Earth, H_2O, N_2, and CO_2 were the most abundant. They offered a nearly inexhaustible supply of hydrogen, carbon, nitrogen, and oxygen. Hence, organic molecules must have been among the dominant products of the chemical reactions. This is the case in interstellar clouds, in which these elements are also abundant and in which organic molecules are manufactured copiously. The same was true of the Sun's protoplanetary disk, as manifested by the presence of organic molecules in meteorites.

Just how readily organic molecules form under conditions similar to those on the early Earth was first demonstrated experimentally by two American chemists. Stanley L. Miller and Harold C. Urey (1893—1981), in the early 1950s, using a glass apparatus. They filled the lower part of the apparatus, which included a small flask, with water to simulate the juvenile

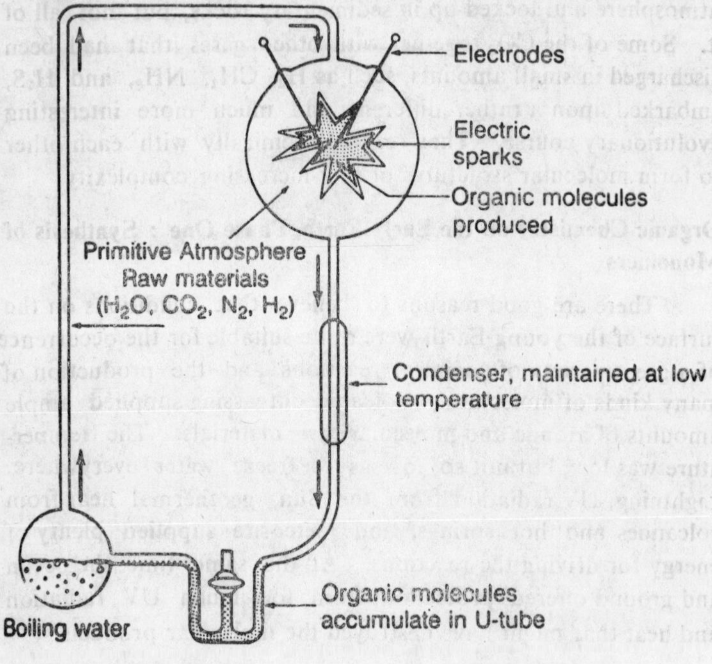

Fig. 2.2. Apparatus used in Miller-Urey experiments.

ocean. They pumped a gaseous mixture of CH_4, NH_3, and H_2 into the upper part to represent the primitive atmosphere. Then they boiled the water in the small flask to produce steam and to drive the gases in a closed circuit through the apparatus. At the same time, they generated electric sparks in the larger upper flask to simulate lightning in the primitive atmosphere and to provide a source of energy for chemical reactions. Below the large flask, the cooled the circulating gases so that the water condensed and returned as droplets to the lower part of the apparatus, thus completing the circuit. After they had run the experiment for a day or two, the water in the lower part of the apparatus turned pink and later it became red. This indicated, and careful analysis confirmed, that organic molecules (such as hydrogen cyanide, formaldehyde, sugars, amino acids, bases) had been formed by the electric sparks and were trapped in the liquid water.

When Miller and Urey first carried out their experiment, it was commonly thought that hydrogen-rich molecules had been the major constituents of the Earth's primitive atmosphere, Hence, they used the gases CH_4 and NH_3 instead of CO_2 and N_2. Since then, the experiment has been repeated many times with various combinations of CH_4, C_2H_4 (ethylene), C_2H_6 (ethane), CO_2, CO, NH_3, N_2, H_2S, H_2, and H_2O and with different sources of energy, from intense heat (from about 1100 to 1600 K) to UV radiation and electric sparks. Some experimenters added sand, clay, or lava rocks to simulate the catalytic effects the ground of the young Earth might have had on the chemical reactions. All of the experiments yielded rich mixtures of organic molecules.

Apparently, the main requirement for synthesizing organic molecules is the presence of hydrogen, carbon, nitrogen, and oxygen in some molecular form or other. The details—whether the atomic raw materials are supplied as H_2O, CO_2, N_2, and H_2, as H_2O, CH_4, NH_3, and H_2, or in some other form—matter little. Nor does it matter much what kind of energy source is used, as long as it is sufficiently intense to drive the chemical reactions. The chief limitations are that oxygen not be present as O_2 and that the newly formed molecules do not remain exposed to the energy sources for too long. If they are, the same energy source the creates them also destroys them. Protection from too much energy is achieved by dissolving the molecular products in liquid water (that is, in the juvenile ocean) or by allowing them to attach themselves to the surfaces of grains of sand, clay, mud, or lava rocks.

Note, however, that the Miller-Urey experiments do not yield many organic molecules of prebiotic importance if the only raw materials are H_2O, CO_2, and N_2. To enhance the production of organic molecules, H_2, CH_4, NH_3, or other hydrogen-rich molecules need to be present among the starting materials, at least in small amounts (without them, the chemical reactions are dominated by the reactivity of oxygen). This is particularly true of the production of hydrogen cyanide (HCN), which is a precursor molecule of many of the more complex organic molecules that are believed to have played significant roles in prebiotic chemistry.

Synthesis of Amino Acids, Sugars, and Bases

How are the simple starting materials converted into organic molecules in experiments simulating primitive Earth environments? Much attention has been paid recently to answering this question. Here are a few examples of these chemical reactions.

When molecules such as H_2O, CO_2, N_2 and H_2 are exposed to electric sparks, intense heat, or UV radiation, their bonds are broken and, temporarily, free atoms (H, C, N, O) and molecular fragments (for example. OH and CO) result. Almost instantaneously, new bonds reform between the atoms and the molecular fragments, but often in combination that are different from the original ones. This breaking and reforming of chemical bonds constitutes the mechanism of chemical reactions by which new kinds of molecules are created. Such chemical reactions take place in statistically predictable ways and according to well-known physical laws. Given identical starting conditions, the molecules produced will, on the average, always turn out the same.

Two of the most common molecules that result during the initial stages of a Miller-Urey experiment are hydrogen cyanide (HCN) and formaldehyde (H_2CO). Both of them are important intermediates in the formation of still more complex organic molecules, in particular of amino acids. For instance, reactions among HCN, H_2CO, and water yield the amino acid glycine. The reactions involve several steps, but they may be summarized by this chemical equation :

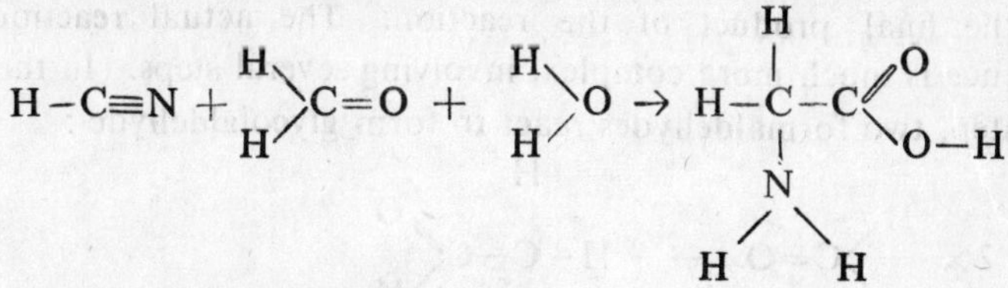

Hydrogen cyanide   Formaldehyde   Water   Glycine

Reactions involving more complex aldehydes than formaldehyde yield more complex amino acids, The nature of the end product depends somewhat on the sources of energy used and on the presence of certain atoms among the starting materials. If electric sparks are used, glycine, alanine, leucine, serine, threonine, asparagine, and other relatively simple amino acids result. If the energy source is heat of about 1600 K, some of the products are amino acids with ring structures, such as phenylalanine and tyrosine. If UV radiation is used and H_2S is present in the gas mixture, small amounts of sulfur-containing amino acids are formed.

In addition to being precursor molecules for the making of amino acids, formaldehyde and hydrogen cyanide may also be the starting materials for the synthesis of other organic molecules. For instance, reactions among five molecules of formaldehyde, in the presence of calcium carbonate, produce a complex mixture of end products, including the five-carbon sugar *ribose*:

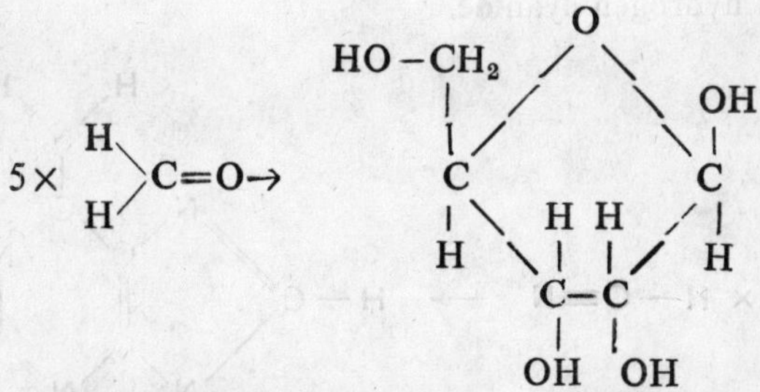

Formaldehyde (× 5)   Ribose

As before, this formula merely summarizes the input material and the final product of the reaction. The actual reaction sequence is much more complex, involving several steps. In the first step, two formaldehydes react to form glycolaldehyde :

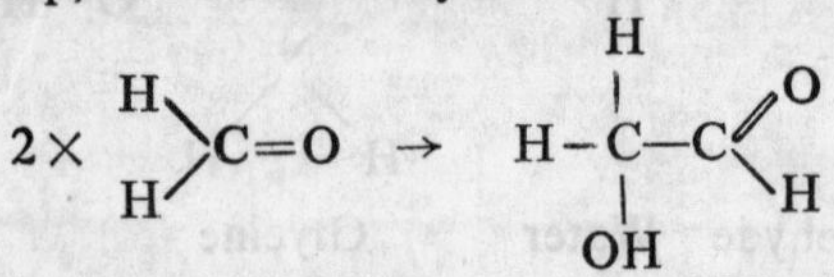

Formaldehyde (×2) Glycolaldehyde

Glycolaldehyde may be either the precursor of ribose, if three more formaldehydes are added (as in the reaction shown above), or the precursor of the amino acid serine, if HCN and H_2O are added. In either case, the reactions are *autocatalytic*. At first very little happens. Then suddenly, after several hours. ribose and serine are produced. Apparently, molecules of glycolaldehyde slowly form during the induction period. Once they exist, they act as catalysts for the production of more glycolaldehyde and then the reactions to ribose and serine run their course rather rapidly. Six-carbon sugars, such as glucose and fructose, are formed by similar reaction pathways.

Not all organic molecules of biological interest are as readily produced as the amino acids and sugars. For example, the syntheses of the bases of DNA and RNA call for rather high concentrations of starting molecules, which on the early Earth probably occurred only under unusual circumstances. *Adenine*, which is the easiest base to make (perhaps that is why it is the most common base found in organisms, being present in RNA, DNA, and ATP), requires the reaction of five molecules of hydrogen cyanide.

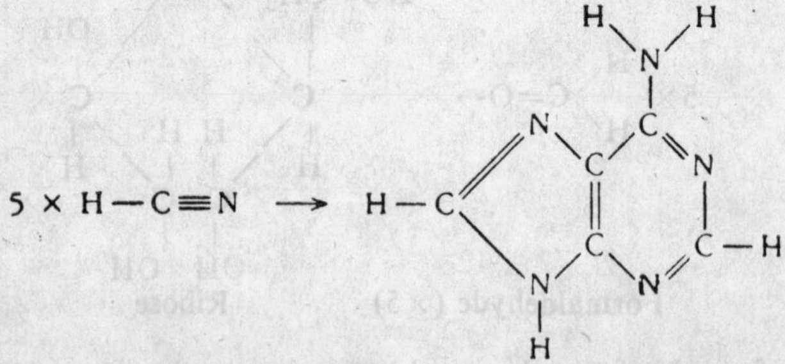

Hydrogen cyanide (×5) **Adenine**

Again, this formula is merely a summary. In reality, several sequential reactions must occur, and the presence of UV radiatian and ammonia are helpful. *Guanine*, the other double-ringed base, is very similar to adenine except that possesses an oxygen atom and its side groups are slightly different. Its synthesis starts out identically to that of adenien but in the final steps of the reaction sequence, water and cyanogen ($N\equiv C-C\equiv N$) or urea ($H_2N-\underset{\overset{\|}{O}}{C}-NH_2$) need to be present.

L-and D-Amino Acids

These have been just a few examples of the numerous chemical reactions that occur in Miller-Urey type and other kinds of experiments simulating prebiotic terrestrial conditions. Many of the molecules produced are identical to those found commonly in present-day life. Others are rather novel and have unusual structures. For instance, many of the amino acids have two configurations, one being the mirror image of the other. In one configuration, the side chain points to the *left* as one looks from the carboxyl group to the amino group. In the other, the side chain points to the *right*. The two configurations are illustrated in Fig. 2.2 for alanine.

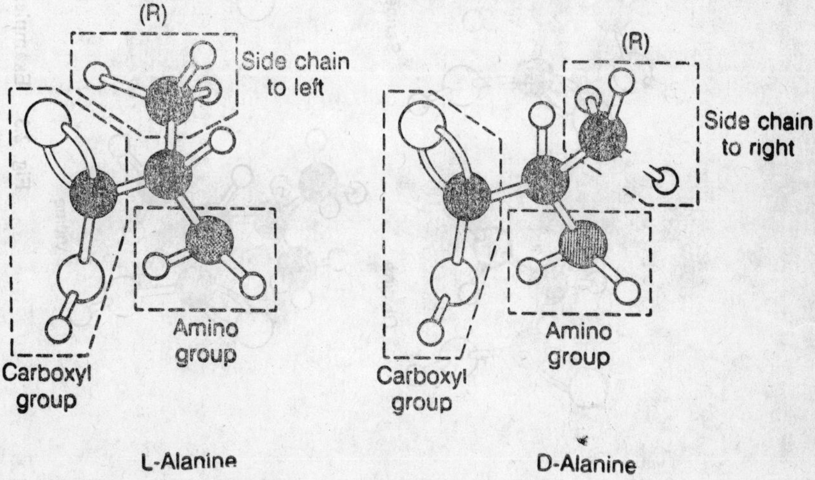

Fig. 2..2 The two configurations of alanine.

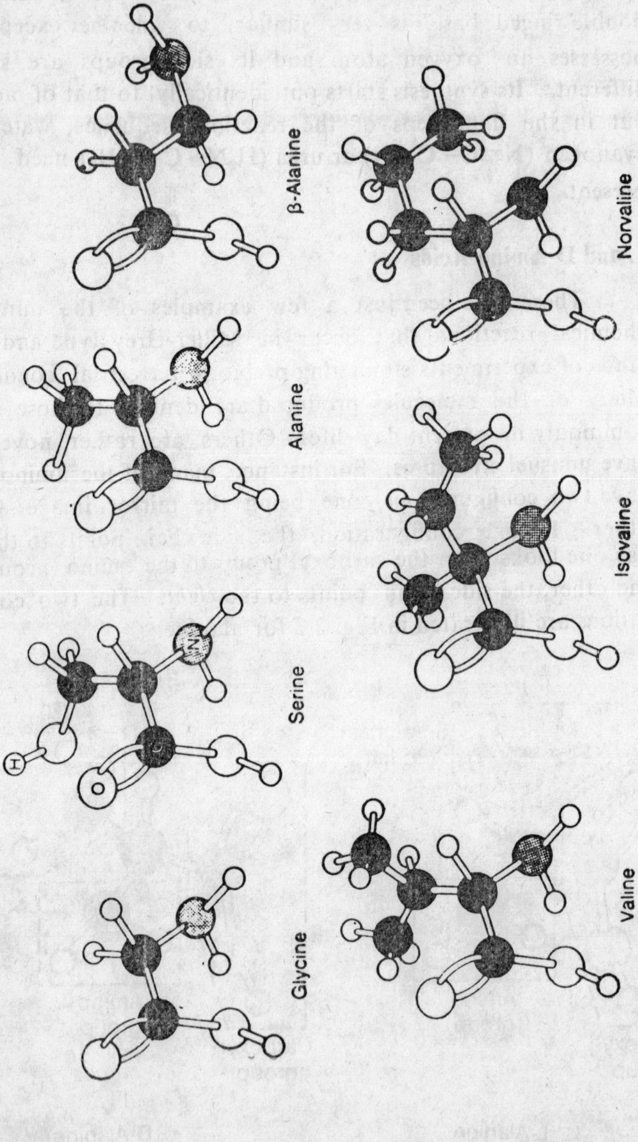

Fig. 2.3. Examples of amino acids produced on the early Earth.

The two kinds of amino acids are known as L- and D-amino acids, where L and D stand for *levo* and *dextro* meaning "left" and "right." Both kinds are found in meteorites, and it is reasonable to assume that they were manufactured by prebiotic terrestrial chemistry as well. In contrast, with few exceptions today's organisms produce only L-amino acids. Somehow, during the origin of life, L-amino acids were selected over D-amino acids. Once that had happened, enzyme-aided biochemical reactions kept manufacturing chiefly the L-types. Was this selection an accident ? Was it due to certain conditions on the early Earth, such as the particular molecular structures of clays and rocks, that may have acted as the first catalysts in prebiotic chemical reactions ? Or do L-amino acids have some inherent advantage in the chemistry of life, perhaps due to some intrinsic bias in physical laws that is not yet understood ? A number of answers have been suggested, but none of them is convincing. We don't really know.

In addition to L- and D-amino acids, other molecular variations that do not occur in present-day life show up in Miller-Urey experiments and, presumably, were also synthesized on the prebiotic Earth. An example is valine, which is produced in three different versions or *isomers*—valine, norvaline, and isovaline—as illustrated in Fig. 2.3. All three of these isomers have the same chemical formula, $C_5H_{11}O_2N$, and they are amino acids, but their side chains are put together differently.

We have seen that varying the conditions—the starting materials, energy sources, and catalysts—of the Miller-Urey experiment and others simulating early-Earth conditions produces a great variety of organic molecules. Some of the molecules form readily under almost any kind of condition as, for example, many of the amino acids. Others, like the bases, require rather special conditions and are more difficult to make (though some of them are found in meteorites).

On the young Earth, environmental conditions must have spanned an enormous range. In the rivers, lakes, and oceans water was amply available, while land parched by the Sun was characterized by dryness. The raw materials varied from molecules without hydrogen. like CO_2 and N_2 to local concentrations of H_2, CH_4, C_2H_6, NH_3, H_2S, and other hydrogen-rich molecules. In the northern and southern latitudes the temperature was well below freezing, while near thermal pools it was in the hundreds of degrees, and near lava flows it was well above 1000 K. Interplanetary debris ranging in size from microscopic dust to bodies kilometers across bombarded the Earth, bringing in new materials and locally heating the atmosphere and ground. The Sun's UV radiation penetrated freely to the ground. Primordial storms produced lightning and released concentrated forms of heat. Rocks and mineral grains of various sizes, textures, and compositions offered additional raw materials, besides those discharged by outgassing, and were available to serve as catalysts for many of the reactions.

From day to night, season to season, and year to year, the early terrestrial conditions fluctuated. We may expect that in the course of thousands to many millions of years a broad range of chemical reactions occurred and that organic molecules were produced in great abundance and in many different forms. Even molecules that are very difficult to make were produced occasionally. The American biochemist, George Wald (1954, 48), put it as follows : "Given so much time, the 'impossible' becomes possible, the possible probable, and the probable virtually certain...... Time is in fact the hero of the plot."

Once the molecules were formed, they were not easily destroyed as they would be today by bacteria and molecular oxygen. The oceans, rivers, and lakes offered protection from UV radiation and excessive heat as did sand, clay, and mud. Thus, organic molecules accumulated until, as Haldane suggested, the primordial ocean reached the consistency of dilute soup. We do not know exactly how concentrated the soup became. That depended on the rate of synthesis of organic molecules as well as on the rate of their destruction. It also depended on the volume of the ancient ocean. However, we may be certain that at least locally, such as in shallow marine basins, tide pools, and lakes, its concentration was sufficiently great for the next phase in chemical evolution to get started – the assembly of simple organic molecules (the monomers) into polymers.

Organic Chemistry on the early Earth,

phase two : Synthesis of Polymers

The prebiotic synthesis of simple organic molecules was only the beginning of a complex and multifaceted evolution of

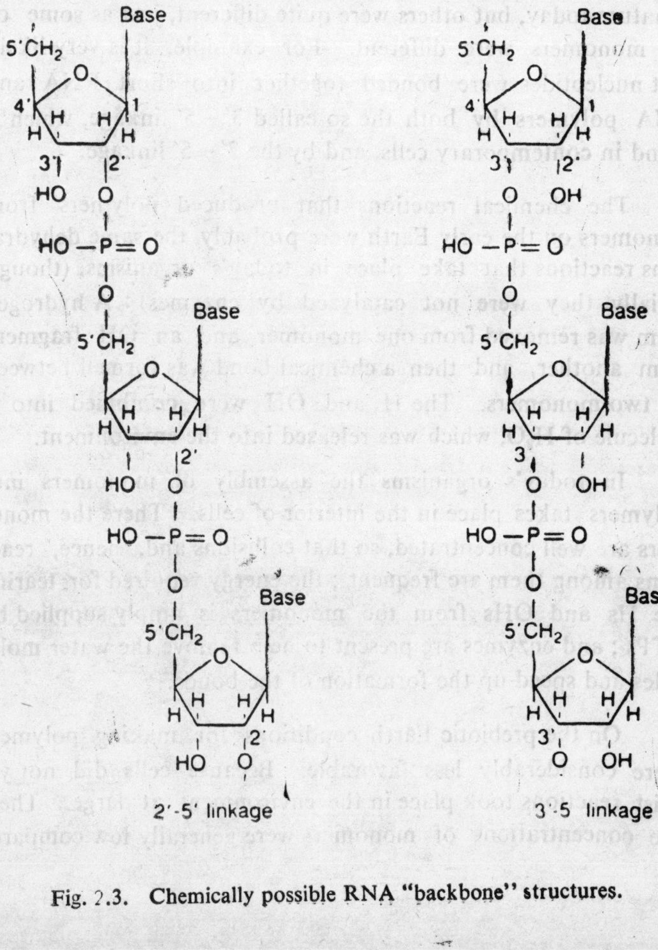

Fig. 2.3. Chemically possible RNA "backbone" structures.

chemical reactions. As amino acids, sugars, bases, and other molecules accumulated and their concentrations in the primordial soup and in the clay, sand, and mud of the young Earth increased, further chemical reactions linked them together into larger molecular structures, namely polymers. Amino acids were joined into peptide chains ; ribose, bases, and phosphates were combined into nucleotides ; and nucleotides were connected into strands of RNA and, possibly, other kinds of nucleic acids. These are just a few examples of polymers that probably were constructed from monomers by prebiotic chemistry. Many early polymers were identical or similar to those found in nature today, but others were quite different, just as some of the monomers were different. For example, it is very likely that nucleotides were bonded together into short RNA and DNA polymers by both the so-called $3'-5'$ linkage, which is found in contemporary cells, and by the $2'-5'$ linkage.

The chemical reactions that produced polymers from monomers on the early Earth were probably the same dehydrations reactions that take place in today's organisms (though initially they were not catalyzed by enzymes) : A hydrogen atom was removed from one monomer and an OH fragment from another, and then a chemical bond was formed between the two monomers. The H and OH were combined into a molecule of H_2O, which was released into the environment.

In today's organisms the assembly of monomers into polymers takes place in the interior of cells. There the monomers are well concentrated, so that collisions and, hence, reactions among them are frequent ; the energy required for tearing the Hs and OHs from the monomers is amply supplied by ATPs ; and enzymes are present to help remove the water molecules and speed up the formation of the bonds.

On the prebiotic Earth conditions for making polymers were considerably less favorable. Because cells did not yet exist, reactions took place in the environment at large. There the concentrations of monomers were generally low compared to the interiors of present-day cells ; energy was available, but most of it not in forms best suited for driving the dehydration reactions ; and enzymes were lacking. In the oceans and lakes the absence of enzymes presented a particularly serious problem. Without enzymes it was difficult to form bonds by removing water molecules from monomers and expelling them into surroundings that already consisted largely of water. The reactions were much more likely to run in the opposite direction, namely in the direction of taking water molecules from the surroundings, adding them to polymers, and splitting their bonds (a process called *hydrolysis*).

Concentration Mechanisms

With all of these obstacles, how were the first polymers assembled ? We may assume that in local areas conditions were occasionally suitable for polymerization. For instance, evaporation of water in shallow lakes and ponds during dry spells might have satisfied the requirement of concentrating the monomers by leaving most of the organic molecules behind on the muddy bottoms. The heat near hot springs or from freshly expelled lava flowing into bodies of water would have had similar effects. The periodic rising and falling of the water level in tide pools would have regularly concentrated the molecules, especially in hot, dry climates. Still another concentration mechanism would have been the freezing of lakes and ponds during winter. As water froze to ice, organic molecules would have become concentrated in the remaining liquid water. (The same method was used by American pioneers in the making of applejack. A barrel of cider was put outside during the freezing weather and left standing until most of the water in the cider was solid ice, leaving the liquid alcohol concentrated as applejack in a small volume near the center of the barrel. The colder the temperature, the higher was the proof and, hence, the alcohol concentration of the applejack.)

The concentration mechanisms just discussed were all a result of changes in the environment. Concentrations may also have been accomplished by the organic molecules themselves.

For instance, in the laboratory when amino acids are put in water that is then heated to above the boiling point, peptides form and cluster spontaneously into spherical, membrane-enclosed structures called *proteinoid microspheres*. These structures are about 1 μm across and, during their self-assembly, concentrate in ther interiors many of the organic molecules present in their vicinity. Another concentration mechanism occurs when certain combinations of organic polymers, such as peptides, carbohydrates, and nucleic acids, as mixed together in water. They spontaneously organize themselves into clusters that are marked off from their environment by structured layers of water molecules. These clusters are called *coacervates*. They have sizes of up to 500 μm and, like the proteinoid micropheres, concentrate organic molecules in their in their interiors. Both proteinoid microspheres and coacervates may well have played significant roles in concentrating organic molecules in the primordial soup of our planet,

Energy Sources

The second requirement for assembling polymers—the availability of Energy—was probably at first satisfied by lightning, heat, and UV radiation, even though these sources of energy are in general not very efficient in linking monomers together. As the number and diversity of monomers increased, the chemical energy stored in the bonds of some of them became available for dehydration reactions as well. Hydrogen cyanide, cyanogen, and other nitrogen-containing molecules similar to cyanogen were probably most effective in driving the reactions. Very likely, chains of phosphate molecules, the *polyphosphates*, were also important energy sources. In fact, polyphosphates may have been the precursors of ATP, which consists of a triphosphate chain attached to an adenine-ribose trunk. Polyphosphates must have been quite abundant on the surface of the early Earth. They would have formed readily by mild heating of minerals containing phosphate, particularly during dry periods when the minerals became concentrated (along with other materials, such as organic molecules) on the bottoms of lakes and ponds.

Catalysts

The third requirement for making polymers is the presence of catalysts for removing water molecules from the monomers and speeding up the reactions. Until enzymes evolved, the most obvious catalysts were the surfaces of mineral grains. Many clays consist of very thin sheets of silicates, that are separated from each other by molecules of water. The water layers would have given the organic monomers easy access to the silicate surfaces, where they could have become attached and been ready to undergo dehydration reactions with newly arriving monomers. Polyphosphates would have been amply available in the minerals as sources of energy. Experiments indicate that peptide chains as long as one hundred amino acids form in the presence of certain clays. The assumption that the surfaces of mineral grains acted as catalysts on the prebiotic Earth is supported by evidence from astrophysics. It appears that many of the molecules (including organic molecules) that are observed in the dark gas and dust clouds of spiral and irregular galaxies are also assembled on the surfaces of dust grains.

Experiments carried out with proteinoid microspheres suggest that they and other accidentally produced protein-like structures might have acted as catalysts as well. However, until the development of genetic information, the sequences in which amino acids were assembled into peptides and protein-like structures remained largely a matter of chance. Therefore, whatever catalytic properties such protein-like structures may have possessed lacked the specificity and efficiency that characterize enzymes in contemporary cells.

This section described some of the processes by which organic monomers are thought to have been assembled into polymers on the early Earth. However, at present scientists are far from understanding the full range of chemical reactions that took place then. No doubt, processes other than those described played important roles in the assembly of polymers as well our knowledge is limited because it is very difficult to duplicate in the laboratory the early terrestrial environment. As noted in the previous section, ther e are a great many variables to consider an enormous variety of chemically possible precursor molecules, a broad range of likely concentration mechanisms, fluctuations in temperature of the environment from below freezing to well above 1000 K, numerous sources of energy and the effects of many different kinds of inorganic materials from clays to rocks, sand, mud, lava, and sediments. Furthermore, experiments of prebiotic chemical evolution must be conducted in closed apparatuses in order to avoid contamination by microorganisms. In contrast, chemical evolution on the early Earth occurred in an open and ever-changing environment consisting of the atmosphere, water, and land. Above all experiments are limited by the factor of time. Prebiotic chemical evolution took place over thousands to many millions of yea rs, a time span that can not be duplicated in the laboratory.

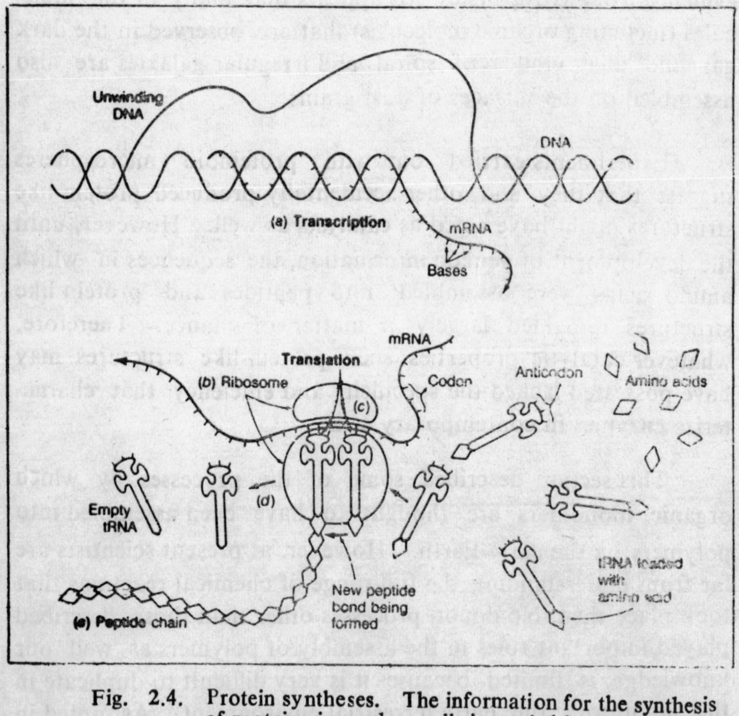

Fig. 2.4. Protein syntheses. The information for the synthesis of proteins in modern cells is carried by molecules of DNA.

Origin of the First Cell

The last two sections described the enormous variety of organic molecules that are believed to have been produced during our planet's early evolution. Some of these molecules were probably similar or identical to those found in life today, while others were quite different. Many survived for long times in sheltered niches, while others were soon broken down into simpler components by the same sources of energy that created them. The simple components were then reassembled into new, more complex molecules. Initially, these chemical reactions occurred in a random, helter-skelter fashion and depended only on the prevailing physical conditions : the concentration of atomic and molecular raw materials, the available sources of energy for driving the reactions, and possibly, the presence of catalysts such as clays, rocks, certain ions, and accidentally produced small protein-like molecules.

Eventually, an *organizing mechanism* or, as Manfred Eigen calls it, an "organizing principle" emerged among the reactions. This organizing mechanism included information and the means of translating that information into chemical function. The emergence of this organizing mechanism was the beginning of a chemical evolution that, in the course of time, led to life. Today, such an organizing mechanism is present in the cells of all orgnisms, including those of our bodies. It consists of *genetic information* and a *biochemical machinery* that translates the information into *enzymes*, which, in turn, make the entire biochemical machinery run.

Before going on to the theory of how the organizing mechanism of life might have arisen, let us review how this mechanism works in today's cells. The compones of the organizing mechanism are the following :

1. *DNA molecules* store *genetic information* by a code consisting of triplets of bases, called *codons*. The codons are arranged sequentially and partitioned into specific units, the *genes*, each of codes for a particular protein molecule.

2. *Messenger RNAs (mRNA)* carry the information of a given gene from DNA to the protein assembly plants, the *ribosomes*.
3. *Transfer RNAs (tRNA)* carry *amino acids* to the ribosomes. For every one of the twenty amino acids found in nature, there exists one or more specific tRNA to which that amino acid becomes attached. Each of the tRNAs is characterized by a distinguishing triplet of bases, called an *anticodon*.
4. *Ribosomal RNAs (rRNA)* are essential components of ribosomes, the protein assembly machineries of the cells.
5. *Enymes* are protein molecules that act as catalysts and are crucial for the smooth functioning of all biochemical reactions, including those of the organizing mechanism. (They function on the molecular level much as tools and machines fuction in a workshop).

The organizing mechanism works as follows:

1. The information of a gene is *transcribed* (copied) from DNA onto mRNA.
2. The mRNA moves to a ribosome, enters and threads its way through it, thus *delivering* the genetic information.
3. As codon of the mRNA pass through the ribosome, tRNAs with matching anticodons bind to the codons and deliver their amino acids, which become linked together by peptide bonds. The now free tRNAs are then released. The genetic information carried by the mRNAs—as a sequence of codons—is thus *translated into a petide chain*.
4. As each peptide chain is assembled, it starts folding back and forth and assumes a unique shape, depending on its sequence of amino acids. The result is a functional protein—a structural protein, a regulatory or transport protein, or an enzyme protein.

In summary, the organizing mechanism of life directs the synthesis of proteins. The proteins—in particular, the enzyme—make the chemical reactions in an organism work, including the chemical reactions of the organizing mechanism itself.

Origin of the Organizing Mechanism

The next question is, "How did the organizing mechanism of life have its start among the disorder and randomness of the chemical reactions on the early Earth ?" Note that the organizing mechanism in contemporary life is based on the presence of both *informational* and *functional* components. The information components are DNA and RNA molecules. The functional components are RNA molecules and proteins (enzymes). Interestingly, the RNAs have dual tasks : they carry information and they perform functions. The DNAs and proteins have single taks each : the DNAs carry information and the proteins perform functions.

There are a number of reasons why in today's cells the RNAs, DNAs, and proteins have these particular features. RNA molecules are usually singlestranded, consisting of a linear sequence of monomers (the nucleotides, which contain the bases A, G, C, and U). The sequential arrangement of monomers allows RNA molecules to carry information, similar to the way DNA molecules carry information. Furthermore, the single strandedness allows RNAs to fold into three-dimensional shapes suitable for carrying out functions such as bringing amino acids to the ribosomes. Some RNAs are also capable of catalyzing certain biochemical reactions and thus act like enzymes, as was recently discovered by a number of researchers. Such catalytic RNAs act either alone or as part of RNA-protein complexes. Finally, short strands of RNA possess autocatalytic properties. The three-dimensional folding of RNA has the additional advantage in that some of the shopes make the molecule resistant to destruction or alteration by hydrolysis and certain other chemical reactions.

In contrast. DNA molecules and proteins, the other components of the organizing mechanism in contemporary cells, do not have the RNAs' dual quality. The DNAs are generally double-stranded and, hence are much less able than RNAs to fold into specific three-dimensional shapes that would allow them to carry out functions. They are mainly used for storing and reproducing genetic information. We may assume, therefore, that DNA was initially not part of the organizing mechanism but was introduced later on. Proteins are made of peptide chains and resemble RNA in that they also fold into a great variety of shapes, which gives their excellent functional qualities. However, peptides are not capable of making faithful replicas of themselves and, hence are not suitable as information carriers.

It is unlikely that all of the components of life's organizing mechanism-RNAs, DNAs, and proteins (enzymes)—came into existence at once in the primordial soup. They must have evolved gradually from primitive and inefficient precursors that were barely distinguishable from accidentally formed molecules. Here it is tempting to ask the old chicken and egg question, as biologists have since the 1930s: "Which came first, the informational or the functional components ?" Experimental evidence gathered by Manfred Eigen and his coworkers, the American biochemist Leslie E. Orgel, and others suggests that neither came first. The informational and functional components of the organizing mechanism originated and evolved together. The start of this evoltion is presumably to be sought among molecules of RNA because, of all the components of the organizing mechanism in contemporary cells, they alone carry information and perform functions.

RNA Quasi-Species

How did RNA molecules become the first component of life's organizing mechanism ? A crucial laboratory experiment for finding an answer was carried out by Leslie Orgel. He demonstrated that short starnds of RNA, such as U—U—U... —U or C–C–C ..—C (poly-U and poly-C strands) are capable of making complementary copies of themselves in solutions containing activated nucleotides (that is, nucleotids with triphosphate chains). For example, poly-C strands form poly-G strands (recall that G is the complementary base of C), and they do so with a fairly high degree of fidelity, meaning that very few errors (the subsiitution of A, U, or C for G, in this example) are introduced. An important feature of this experiment is that the RNA strands have autocatalytic qualities and their replication proceeds without enzymes. This adds realism because, initially enzymes would either not have been available in the primordial, soup or, if present, would have been of poor and nonspecific quality. If zinc ions are added to the solution, much longer strands (up to 40 nucleotides in length) are copied and with greater fidelity than without those ions. The zinc ions act as inorganic catalysts, which, very probably, were available in the soup. Interestingly, today's RNA polymerases (enymes that catalyze the formation of RNA) all contain zinc ions, which led Eigen (1981, 101) to ask, "Has nature perhaps 'remembered' how replication started ?"

Orgel's experiment demonstrated that strands of RNA are capable of self-replication : They provide both the information and the function for making co nplementary copies of the themselves. This result, as the dual roles RNA plays in contemporary cells, suggests that first components in the evolution of life's organizing mechanism were indeed molecules of RNA or, at least, molecules similar to RNA.

We may imagine that life's orgnizing mechanism get its start when short strands of RNA—perhaps only two or three nucleotides long—were randomly assembled in the primordial soup. The original RNAs then began to replicate themselves by processes such as those in Orgel's experiment. In the absence of well-developed enzymes, errors were commonly made during replication, despite the RNAs' short lengths. Hence, the replicated copies differed frequently in minor and major ways from the original ones with regard to sequences of bases and lenghts. *Variation*—the first requirement of Darwinian evolution—was thus introduced right from the beginning into the pool of RNAs in the primordial soup.

Along with creation of variation among the RNAs, *natural*

selection – the second requirement of Darwinian evolution – got its start as well. The different kinds of RNA were not all equally successful in replicating themselvs, avoiding errors during replication, and surviving. For example, RNAs with an abundance of Cs and Gs probably tended to replicate themselves more faithfully than those with mostly Us and As.; and RNAs with certain kinds of three-dimensional folding were more resistant to being broken up by water molecules and other chemically reactive compunds. It probably was also at this early stage that the particular bonding between nucleotides that we and universally in nature today out over other, competing bonds.

In the course of time, the most successful RNA strands accumulated and gave rise to so-called *quasi-species* of RNA with variations in characteristics among their member RNAs, analogous to the variations observed today among their member of species of organisms. The quasi-species of RNA differed from one environment to another ; and as the environments change, the quasi-species either changed also and adapted or they disappeared. Thus Darwinian evolution, based on variation and natural selection, appeared in the primordial soup long before life as we know it today had come into existence.

Very likely, the chemical reactions by which the earliest strands of RNA replicated themselves were aided by inorganic cataysts such as clays, lava, rocks, sand, and ions of zinc and other elements. Accidentally produced protein structures such as Fox's proteinoid microspheres might have placed a role as well.

In the absence of specific enzymes that catalyzed replication, Eigen (1981, 103) estimates that the early RNAs could at most have been about 50 to 100 nuclecotides long, which is comparable to the lengths of today's tRNAs. Had they been much longer, the number of errors introduced during replication would have been so great that within a few generations their particular base sequences would have become changed beyond the range of viability.

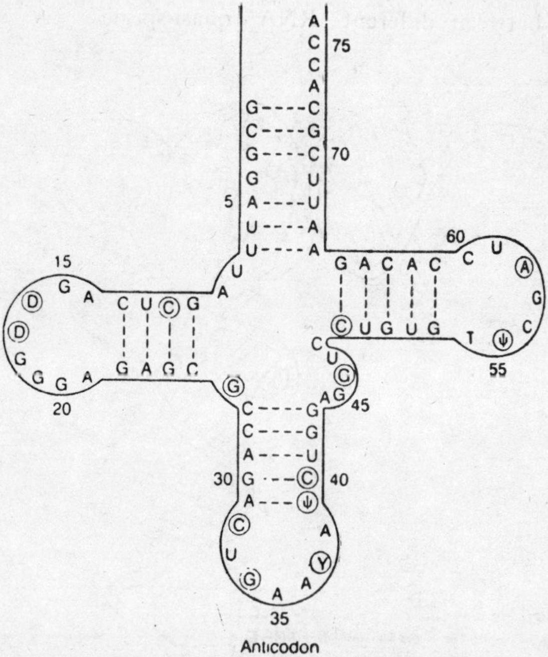

Fig. 2.5. Two-dimensional representation of the nucleotide sequence (76 bases) of yeast tRNA (coded for phenylalanine).

RNA Hypercycles

As long as the strands of RNA consisted of no more than 100 nucleotides, they could do little else than help in their own replication. They were not lor g enough to store information for the assembly of enzyme catalysts. However. enzymes were needed to allow the organizing mechanism to progress further. They were needed for the faithful replication of longer RNAs, and the longer RNAs were needed to store the information for the assembly of the enyzmes. Furthermore, specialized functional RNAs were needed, analogous to today's tRNAs (and possibly rRNAs) to help in the translation of the information into enzyme structure.

According to theoretical studies by Eigen and his cowor-kers (1981, 104-107) the evclution tcward longer strands of

RNA and enzymes required the development of cooperative coupling between different RNA quasi-species. A simple

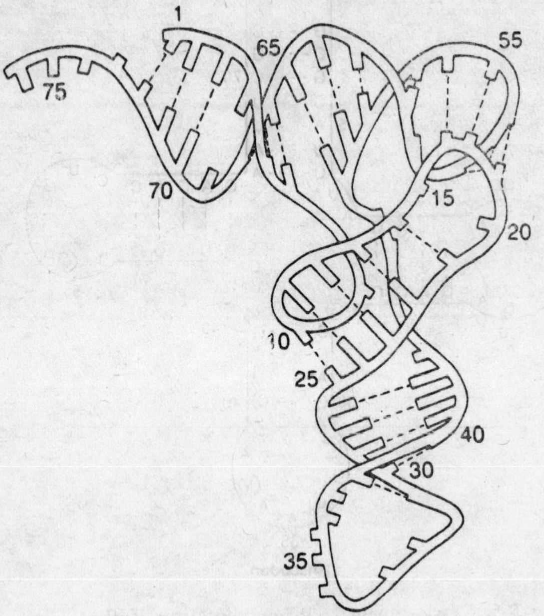

Fig. 2.6. Model of the three-dimensional folding of yeast RNA. The ribosephosphate backbone is drawn as a continuous structure; positions of the bases are shown as bars; and the H-bond between the bases are indicated by dashed lines.

cxample of such couplings involves three quasi species. The RNAs of one of the quasi-species carry information for the assembly of enzmes (albeit, very primitive ones) and the RNAs of the other two quasi-species function as tRNAs. With the aid of the tRNAs, the information of the first quasi-species is translated into primitive enzymes. The enzymes, in turn, aid in the replication of the RNAs of all three quasi-species. Such cooperative couplings are called *hypercycles.*

In reality, hypercycles in the primordial soup probably were never as simple as the one just described. We may assume that many more quasi-species of RNAs and kinds of enzymes than those indicated in the example were required, even for inefficient hypercycles. Hypercycles probably evolved from single RNA quasi-species that consisted of large numbers of RNAs with the usual variations with regard to length and base sequence. At first the participating tRNAs, helping in the assembly of peptides according to information carried by the longer RNAs, were quite inefficient. Likewise, the peptide chains, which functioned as enzymes, were quite inefficient also. They were probably not much more effective in this task than

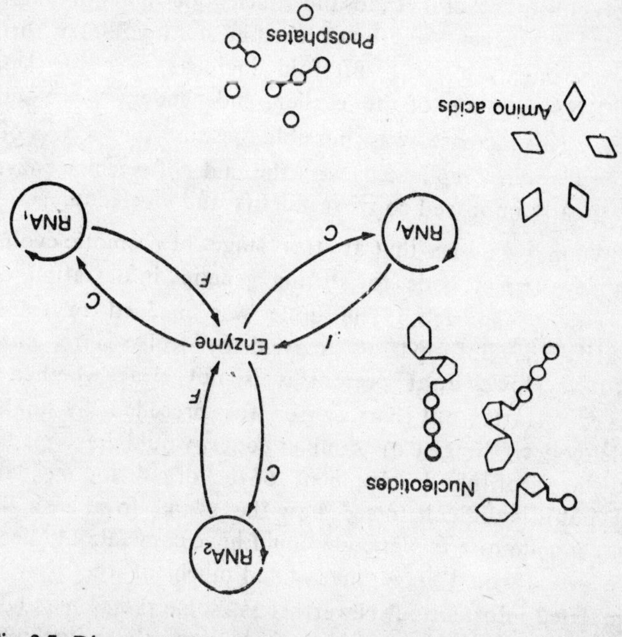

Fig. 2.7. Diagrammatic representation of a hypercycle, consisting of three distinct RNA quasi-species (labeled RNA/ RNA_1,) RNA_2, and enzmes.

many of the accidentally produced peptide chains and other inorganic catalysts. In the course of time, however, the lengths of some of the participating RNAs increased, and their information content—coding for the assembly of enzymes—grew more distinct. Other RNAs remained short (50 to 100 nucleotides) and took on the role of today's tRNAs. The enzymes thus produced became more specific and efficient in their tasks. Slowly the original RNA quasi-species evolved into a number of distinct RNA quasi-species, with each quasi-species specializing in a particular task: to store information for the assembly of enzymes or to function as tRNAs. Thus the different RNA quasi-species, along with their enzymes, became firmly coupled and dependent on each other for survival. They evolved into hypercycles.

The longest RNAs of hypercyclic quasi-species were conceivably up to several thousand nucleotide units in length, as suggested by the gene lengths of certain contemporary viruses. That was roughly a ten-to-fiftyfold increase over the lengths of the RNA strands of the earlier, independent RNA quasi-species. This increase was possible because the RNAs of the hypercycles were replicated with the aid of specific enzymes, which greatly improved copying fidelity and efficiency.

We may assume that at this stage of prebiotic evolution the modern triplet code for storing genetic information came into existence as well. The code was needed to translate accurately the information carried by RNAs into enzyme structure. However, at present it is not clear whether the triplet code started out that way or was preceded by singlet or doublet codes. Singlet or doublet codes would have made the initial stages of the development of the organizing mechanism easier, but it is difficult to see how the switch from a singlet or a doublet code to a triplet code could have come about, because such a switch would have invalidated much of the previously accumulated information. Nevertheless, some researchers believe that the first code was singlet or doublet and that somehow, at a later stage during the origin of life, it became converted into the triplet code that is in universal use today.

Prebiotic evolution progressed from independent RNA quasi-species to hypercycles not only because the latter produced their own enzymes. It also progressed in this direction because, through the enzymes, the participating RNA quasi-species became dependent on each other and were forced to cooperate for the common good. They could not afford to outcompete and to eliminate each other. To survive, the had to cooperate.

This cooperation greatly increased their reproduction rate and made them the dominant quasi-species in the soup.

The advantages of hypercyclic cooperation may be illustrated by an example from economics. Let us imagine a primitive human society in which every family is self-sufficient and competes with other families for available raw materials. Each family grows its own food, makes its own tools, builds its own shelters, and defends itself against competitors. Such an economy may be compared with the earlier, noncooperating RNA quasi-species, which competed with each other for the available raw materials. A much more efficient and productive economy results when the families begin to join into cooperative units, with each family specializing in specific tasks. Some families are farmers and grow food. Others are blacksmiths, butchers, bakers, or carpenters and specialize in the making of tools, slaughtering, baking, and building. Such a cooperative economy will easily outproduce and outcompete the simpler one in which each family fends for itself. The same was true of the hypercycles of coupled and cooperating RNA quasi-species, in which the different kinds of participating RNAs had specialized informational and functional task and in which they cooperated for the common good.

Another example of hypercycles is our present-day ecosystem, with its many interdependent and cooperative linkages among animals, plants and microorganisms. Without the support from microorganisms, animals and plants could not exist. Likewise, many microorganisms depend on plants and animals for their nutrients. Furthermore, all animals depend (directly or indirectly) on plants for food and oxygen. In analogy, we may regard the RNA quasi-species that were linked into cooperating hypercycles in the primordial soup as our planet's first ecoysytem.

Protocells

The arrival of hypercycles of coupled and cooperating RNA quasi-species constituted an important step forward in the development of life's organizing mechanism. However, there still remained a serious obstacle. There still existed no decisive feedback from the enzmes to the information content of the RNAs. Natural selection did not favor RNAs because they carried information for superior enzymes; it favored RNAs that were the most stable and reproduced most rapidly under the prevailing conditions, regardless of whether they also carried useful genetic information or not. Even RNAs that carried no information at all and performed no functional task benefited from the enzymes produced by other RNAs.

This feedback problem resulted from the fact that up to now all chemical reactions had taken place in the open environment of the primordial soup. As long as this remained the case, the obstacle could not be overcome. Only by enclosing the hypercycles with membranes and forming protocells did evolution progress further. Competition could then occur between the protocells, which forced natural selection to focus on the fitness on the protocells' entire-enzyme, driven biochemical. Cells that survived possessed, by definition, the most advantageous enzymes, and (because the cells survived) their genes were passed on to the next generation. Thus, the feedback from enzyme to gene or from phenotype to genotype--that is characteristic of all contemporary life became established.

Besides solving the feedback problem, membrane enclosure had other advantages. It kept the enzymes produced by the hypercycles close to the RNAs and prevented their diffusion into the environment. It also created the possibility of concentrating biochemical raw materials—such as amino acids, nucleotides, and phosphates—by the development of elective transport mechanisms into the cells' interior.

At present no one knows just how the evolution from hypercycles to protocells came about. It was a step that probably required the parallel evolution of numerous major and interconnected biochemical capabilities; the evolution of suitable membranes; selective transport of the right kinds of raw materials through the membranes into the cell interiors; elimination of waste products; extraction of energy from energy-rich molecules; replication of genetic information with high fidelity; and efficient translation of the genetie information into enzymes and other kinds of proteins.

Though we do not know how these capabilities arose, it is clear that to function reasonably efficiently protocells required many more enzymes than were produced by the earlier hyper-cycles. Furthermore, these enzmes needed to be constructed with an ever higher specificity for catalyzing particular chemical reactions. Eventually, the information required for assembling the enzymes became so great that it could not have been stored and passed on from generation to generation if strands of RNA had remained the only information carriers. Too many errors would have been made during the replication of RNA, despite the assitance by enzymes, and any one of the errors might have been fatal to the cells. The only way out was the development of an error suppression mechanism.

The error suppression mechanism that evolved depended on the introdcution of a double-stranded information carrier. That information carrier was DNA. As noted before, the two strands of DNA are complementary—the base adenine (A) of one strand is always bonded to thymine (T) on the other strand, and cytosine (C) is always bonded to guanine (G). Hence, the two strands carry identical information. The presence of the same information twice permitted the development of a "proof-reading" mechanism during replication. While a copy is made of one of the strands of DNA, it is checked by proofreading enzymes against the other, complementary strand. Any errors detected are corrected.

This *error suppression* in all contemporary cells. It must also have evolved, at least in primitive form, in the protocells. It allowed their DNAs to store genetic information that extended over hundreds of thousands to millinos of nucleotides, which probably was enough for the protocells' genetic requirements. The DNAs were replicated sufficiently faithfully that errors—that is, mutations—did not, in general, accumulate to harmful levels. The occasional errors that did occur contributed to the genetic variations of the protocell species and, hence, to their evolution.

Very likely, species of protocells arose separately on numerous occasions in the primordial soup. Nevertheless, the universality today of the structures of proteins and nucleic acids, the triplet code, the use of ATP, and the workings of the organizing mechanism suggest that competition and natural selection eliminated all but one of the early species of cells. The surviving species became the ancestors of the subsequent terrestrial life—from bacteria to protists, fungi, plants, and animals.

There are good reasons to believe the earliest cells were prokaryotes and that they reproduced by binary cell division, similar to the reproduction of contemporary bacteria. Furthermore, they probably derived their energy by fermenting energy-rich molecules, such as sugars, that they found in the soup.

No one knows how much time elapsed before chemical evolution made the transition to the first species of cells. It may have been a million years or less. Or it may have been many hundreds of millions of years. All we know, from the fossil record, is that by about 3.5 billlion years ago single-celled life inhabited shallow areas of the sea. There are also indications that these cells were capable of photosynthesis and, hence. had already evolved considerably beyond the state of the earliest fermenters. For lack of more specific information, let us assume then, somewhat arbitrarily, that protocells arose approximately 4 billion years ago. This is a round figure that will suffice for even though it may be off by a few hundred million years one way or the other.

With the arrival of protocells in the primordial soup, the fundamental components of life as we know them today had come into existence, at least in rudimentary form; the organizing mechanism, consisting of DNAs, RNAs, and enzymes; the triplet code for writing genetic information; the means of selective absorption of raw materials and of expulsion of waste products across the cells' enclosing membranes; the extraction of energy-rich molecules; and the capacity for cell reproduction and dassing genetic information on to the next generation.

The driving mechanism in the development of the protocell was Darwinian evolution. According to the theory presented here, this evolution began with the self-replication of short strands of RNA among the otherwise random chemical reactions in the primordial soup and progressed step-by-step to RNA quasi species, hypercycles, and membrane enclosure of some of the hypercycles. Remember that theory is based on many assumptions and contains may gaps. Some of the assumptions may turn out to be wrong. In fact, researchers in a number of laboratories today are developing and testing alternative theories, some of which rely on RNA and RNA-like molecules as the first replicator and other that don't. Clearly, investigation of the origin of life is at present characterized by great uncertainties and enormous intellectual challenges. Nevertheless it is an exceptionally exciting field of scientific research, dealing as it does with one of nature's most fundamental secrets.

3

The Climatic Environment

Introduction

The patterns of life are influenced by many processes operating at different spatial and temporal scales. Continental drift is recognized as a geologic process that occurs on a global scale over the course of millions of years. Glaciation is a meteorological phenomenon that also has geological impact, but operates in cycles of tens and hundreds of thousands of years. This chapter will focus on general meteorological phenomena of The Earth : weather and climate. Weather is the current condition of the atmosphere. Weather has diurnal and annual cycles and has an immediate impact on the Earth's organisms. Climate is the long-term average of weather. Current practice in the United States describes climatic norms by using a 30-year mean which is updated each decade.

The patterns of weather and climate are an essential ingredient in the broad outline of the Earth's current pattern of life. To set the stage for understanding the underlying dynamic events, this chapter begins by examining the large-scale circulation of the atmosphere and ocean. Then, the focus turns to the relationship of biota and climate, the influence that dynamic patterns of weather have on biota, and the history of climatic change and its influence on vegetation during the past several thousand years.

Meteorological processes

The opration of the Earth's climate depends on the physical properties of water. Water is a versatile substance. It has a high specific heat capacity, and therefore requires much energy per unit mass to increase temperature. Vice versa, water also gives up much energy per unit mass when it cools. At the temperature range of the Earth, water occurs in the three states of matter ; solid, liquid. and vapor. If the average temperature of the Earth were only a few degrees higher or lower the Earth would either support more vapor (humidity) or more ice (glaciers) than it does now.

The relationship between the different physical states of water follows well-understood principles common to all matter. The change from one state of water to another is accompanied by either a gain or a loss of energy. Changes in one state of water to another are not, however, reflected only by changes in temperature. Temperature is a poor measure of energy. For example, it takes only one calorie (4.184 J) of heat to raise the temperature of 1 g of pure water from 18 to 19 °C at standard pressure (760 mb). However, pure water requires an addition of 540 calories (2.259 J) to change from a liquid state at 99 °C to a vapor state at 100 °C at standard pressure (760 mb) (i.e., one calorie (about 4 joules) for a change from 99 °C to 100 °C plus 539 calories (2,255 J) as latent heat of vaporization). The relationships are much more complicated at the Earth's surface because variation of pressure, temperature, the quality of water, and surface albedo all influence vaporization and condensation. Also, the Earth's average air temperature is about 18 °C (54 °F). Consequently, the heat required to vaporize water at 18 °C is about 600 calories (2,500 J) per milliliter of water. Changes of state of water, especially gas to liquid and vice versa, provide a mechanism for transferring energy from one place to another on the Earth. These mechanisms directly influence the pattern of living organism on the Earth.

Heat transfer on the Earth

The yearly receipt of solar radiation differs at various

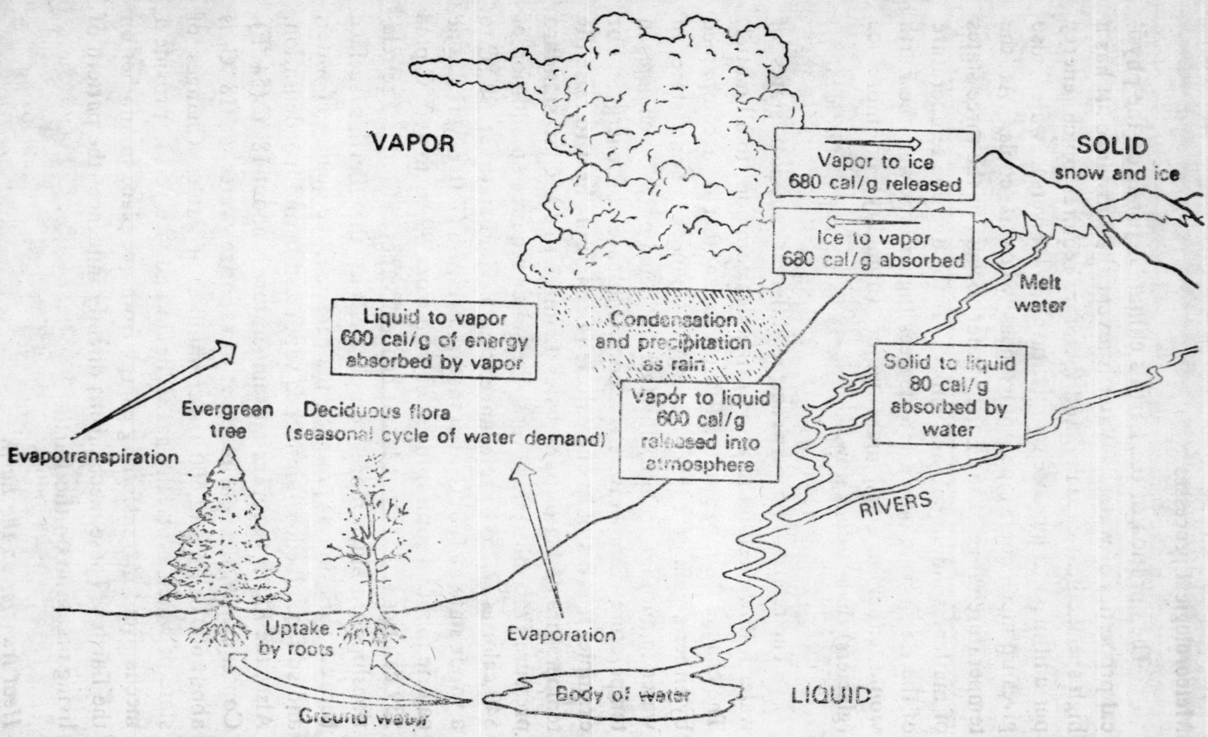

Fig. 3.1. Approximate energy exchanges of water as it changes state at the average pressure and temperature of the Earth's lower atmosphere.

latitudes of the Earth. There is a mean solar radiation surplus (up to 0.12 cal/cm²) at latitudes below 30° and a mean radiation deficit (up to —0.08 cal/cm²) or—0.335 J/cm²) at latitudes above 30°. There are two basic reasons for this radiation difference. First, the solar input is almost constant at the Equator, that is, about 12 hours per day, while above 66.5° latitude, solar input varies from 24 hours per day during summer months to zero hours per day during winter months. Secondly, the intensity of solar radiation reaching the surface differs markedly between the Equator and poles. Intensity of light at different latitudes is related to the geometric aspects of the in coming rediation and the angle at which it hits different surfaces of the Earth's sphere. As the incoming solar energy hits the atmosphere and the Earth's surface at a perpendicular angle at the Equator and at an acute angle at the poles. The difference in net radiation between the Equator and poles is important to the Earth because it sets up the pattern of potential energy which ultimately drives the transfer of fluids and energy on the Earth's surface. Water, especially as it changes state, is the major working fluid for energy transfer and moderation of surface temperatures of the Earth's surface.

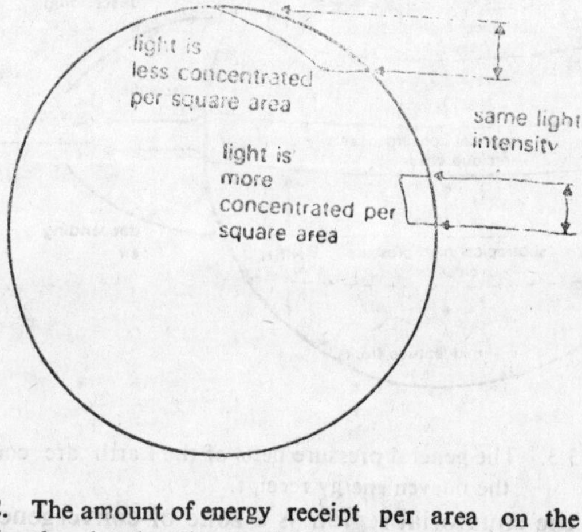

Fig. 3.2. The amount of energy receipt per area on the Earth's surface is directly related to degree of latitude. As the angle of receipt decreases, the energy per area also decreases. Topography or the angle of slope of a particular hill also influences local energy receipt.

The pattern by which fluids and heat energy are transferred is illustrated in the simplified and exaggerated profile of the Earth's atmosphere. The cells of rising and descending air and the meridianal flows (i.e., along lines of longitude) between the Equator and the poles. The major region of surface heating on the Earth occurs in the generally cloudless region of the subtropical latitudes around 15—30° N and S. This is due to the fact that at that latitudes descending air creates predominantly high pressure conditions at the surface. The temperature of the descending air tends to increase as it becomes compressed. Air undergoing temperature increases does not favor the condensation of water vapor. Hence, the subtropical latitudes are zones of relatively low frequency of cloud cover and precipitation. On land surfaces in the subtropical latitudes, evaporation so overwhelms precipitation that liquid water is generally scarce. Without cloud cover or water to evaporate, there is intense daily heating and intense nocturnal cooling of the subtropical land surfaces.

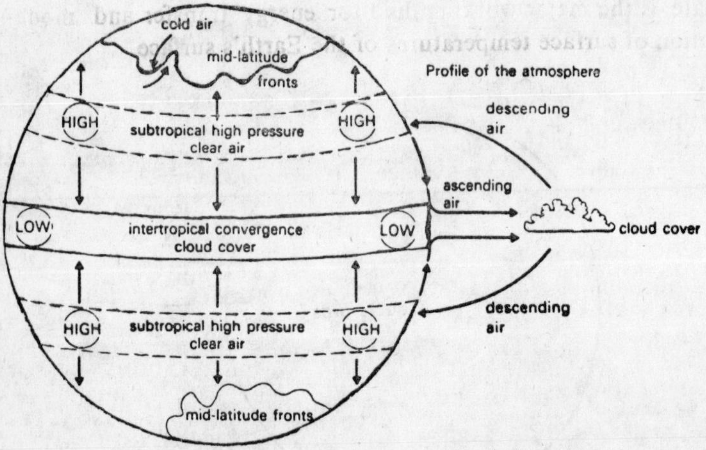

Fig. 3.3. The general pressure belts of the Earth are controlled by the uneven energy receipt.

In the equatorial region is a zone of convergence of subtropical air from both hemispheres known as the intertropical convergence (ITC). The collision of moisture-laden air at the surface forces equatorial air to rise. As it rises it cools and condenses. This process creates an equatorial zone which has a high frequency of cloud cover. The clouds block insolation and the condensation yields large volumes of precipitation. Evaporative cooling has a moderating influence on temperatures. Through these processes, surface temperatures are moderated at equatorial locations.

In mid-latitude regions, a third zone exists in which air flowing poleward from the subtropical zones collides with cool and inherently dry polar air. This action produces another major zone or "front" of cloud formation and precipitation. The mid-latitude zone is often referred to as "temperate." However, as a region of extreme sensonal temperature change it is anything but "temperate." The name temperate is a remnant of the ancient Greek climate classification where to the north of Greece was the frigid climate of Central Europe and to the south of Greece was the torrid climate of the North African desert. Coastal Greece was the site of the original temperate climate. Mid-latitude will be used interchangeably with "temperate" to designate the zone of greatest annual temperature range. Polar air is to cold to hold much water vapor and thus polar areas tend to have very low rainfall, although they do have extensive cloud cover. This general description of the latitudinal climate zones requires several modifications in order to introduce some realistic idea of the dynamic condition of the Earth's weather and climate.

Coriolis Effects

Converting the global heat transfer patterns into the generalized wind patterns on the Earth requires the inclusion of Coriolis Effect (Coriolis 1835). The Coriolis Effect refers to the tendency of moving matter to follow a curved path as a result of the spin of the Earth on its axis. The effect is greatest at high latitudes. In the Northern Hemisphere, water and air flow to the right of their initial direction of movement. In the Southern Hemisphere, water and air tend to flow to the left of their initial direction of movement.

The initial movement of converging ascending air at the Equator and descending diverging air at 30° N and S latitude becomes converted into moving air that is deflected into a circular pattern on the Earth. These winds act in turn upon the oceans to produce currents. Through wind production, about 1% of the incoming solar energy is converted into waves and currents. Most of the wind energy goes into the production of waves. Only a small fraction of the wind energy actually moves water. Due to the geometry of the Earth and the vastness of the ocean, small movements of water are combined, to create locally strong currents.

The Ekman Spiral and the atmosphere ocean interface

The direction of the overall flow of water deviates about 90° from the direction of the wind through a mechanism described by V.W. Ekman in 1902 and known as the Ekman Spiral (Stewart 1969). The Ekman Spiral is explained as follows. The ocean can be envisioned as being made up of a series of independent horizontal layers of water. From the surface to increasing depths, each layer drags each successive layer along at decreasing velocities. In addition, each layer becomes deflected from the direction of the layer above according to the Coriolis Effect, i.e., clockwise (Northern Hemisphere) or counterclockwise (Southern hemisphere). Thus in the Northern Hemisphere, the winds tend to be westerly, but the surface water moves slightly to the right of the wind direction and each successive layer moves to the right of the one above it. One effect is that icebergs move from 20° to 45° off direction of the wind. The larger the iceberg the greater the Ekman effect on it and the farther it moves from the pole. More significantly for the Earth's organisms, the Ekman Spiral mechanism leads to warm currents off east coasts and upwelling cold water along west coasts of continents.

Due to the Ekman Spiral mechanism, the surface water moves from the poles of both hemispheres toward the Equator. A massive quantity of water converges at the Equator and flows westward. The flow of water becomes concentrated into strong current as it is deflected upon the east coasts of the continents.

The Gulf Stream and Kuroshio Current illustrate this phenomenon.

As a result of the heat-holding capacity of water, ocean currents are a major contributor to the transport of heat on the Earth. These currents deliver warm tropical waters into mid and upper latitudes. They also provide an active means of transporting aquatic and terrestrial organism from one site to another. Of special significance for sea life are regions of the ocean at mid and high latitudes where warm currents meet cold water. The collision of masses of differing bodies of water creats sites of upwelling water. The upwelling bottom waters tend to be enriched with nutrients and promote the growth of plankton.

On the west coasts of continents in mid-latitude regions, winds move along the continent from the poles toward the Equator, Again, due to the Ekman Spiral mechanism, the mass of water flows at right angles from the wind direction, or away from the continents. Upwelling occurs along the coast as cold deep water replaces surface water which has been displaced. The pattern of ocean currents and upwelling provides important keys to understanding the distribution pattern of the Earth's organisms.

The Earth's climate is even more complicated. For example, although wind patterns directly drive ocean currents, the distribution of massive quantities of water at various temperatures has a direct effect on the generation of winds. The ocean and atmosphere must be considered as an interlinked system.

An added element of complexity is due to the vertical mixing of the ocean. Temperature and salinity differences of sea water are linked to climate and affect the density and hence, vertical movement of sea water. Cold water is denser than warm water, and highly saline water is denser than water with a lower salinity. There are several general combinations. In subtropical regions of the Earth, salinity increases because of aevporation. However, in polar regions, salinity increases

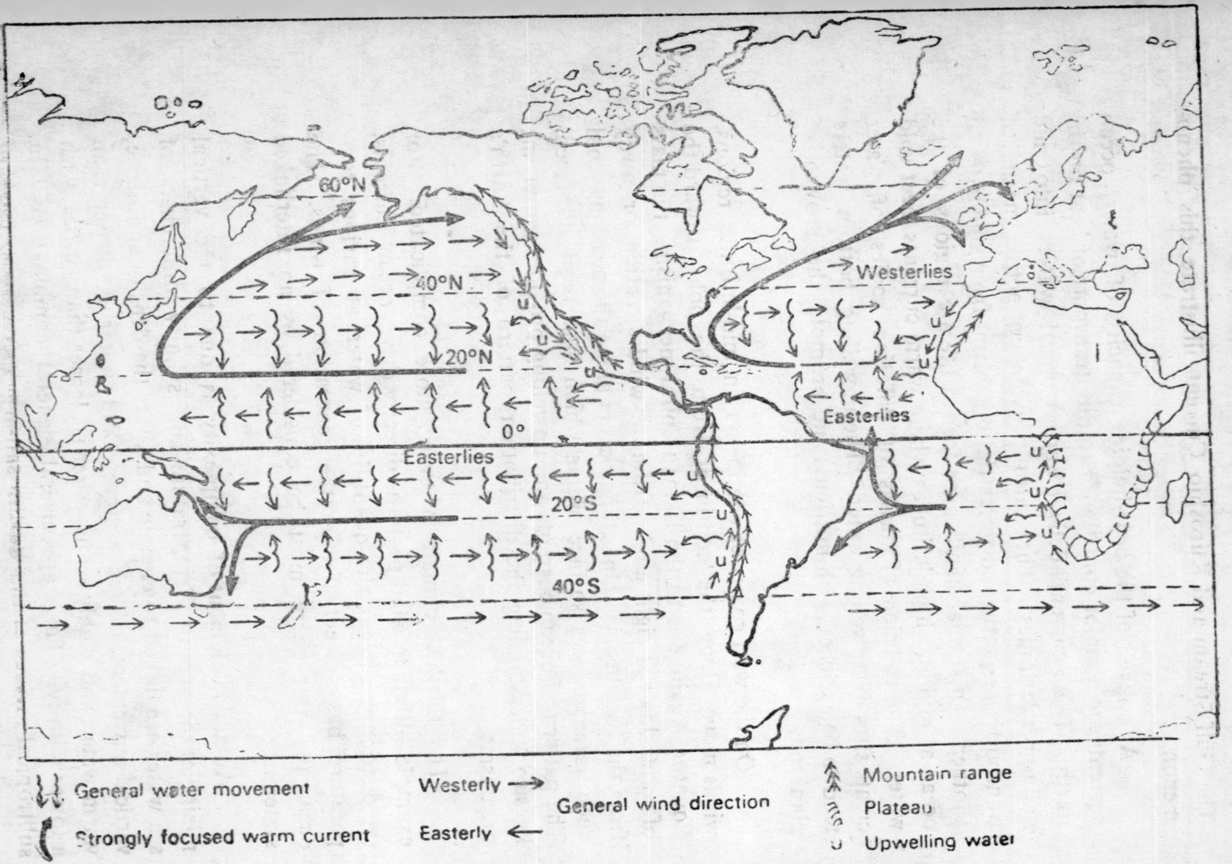

Fig. 3.4. Generalized patterns of wind and ocean current patterns of the Earth.

during winter months due to the removal of salts from ice (or sublimation) as sea water freezes. In humid regions of the Earth salinity tends to decrease because of rainfall and fresh-water runoff from the land. In polar regions, summer melt water from glaciers and ice shelves decreases salinity.

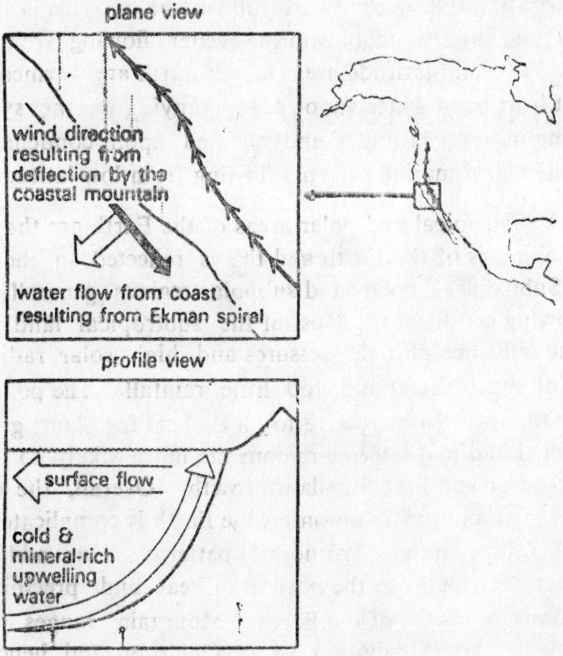

Fig. 3.5. Details of the Ekman Spiral and upwelling water phenomena along west coasts.

The various thermal and salinity, or thermal-haline. combinations produce vertical movements of sea water. The densest sea water displaces the bottom water and causes vertical mixing. The cold saline sea water produced by sublimation in the polar regions is the densest sea water of all. It flows along the continental shelf and settles on the sea floor. The bottom water is replaced at a rate which depends on the volume of the particular ocean. The mixing time of the Pacific bottom water is between 1,000 and 1,600 years, while the mixing time for the

Indian and Atlantic Oceans is about 500 to 800 years (Anikouchine and Sternberg, 1973).

Overall, the climatic effect of the various air flows and marine currents is to moderate the temperatures of large areas of the Earth. The tropics are cooled by the cloudiness, rainfall, and evaporation that occur as a result of the intertropical convergence and by the cool marine water flowing from high latitudes. The mid-latitude areas of the Earth are warmed both by latent heat from water vapor accompanying weather systems originating in lower latitudes and released upon condensation, and by the warm marine currents flowing from low latitudes.

The subtropical and polar areas of the Earth are the most extreme climates of the Earth and this is reflected in the plant cover. Subtropical, polar, and subpolar regions generally have poor growing conditions. Most of the subtropical land areas, under the influence of high pressures and high solar radiation, receive too much heat and too little rainfall. The polar and subpolar tundra regions receive too little heat for plant growth. The tropical and mid-latitude regions are more likely to obtain sufficient water and heat for plant growth. Overall, the distribution of heat and precipitation on the Earth is complicated and does not follow simple latitudinal patterns. One additional factor that contributes to the pattern of heat and precipitation is the mountain ranges of the Earth. Mountain ranges play a major role in the distribution of precipitation and hence the pattern of life on the Earth.

Mountain Climates and Biogeography

The general latitudinal pattern of the Earth's sea-level climates is described. The distribution of mountain topography contributes significantly to the climatic patterns of many areas of the Earth by creating barriers and deflectors of surface air flow.

Orographic rainfall and rain shadow effect

Mountains serve the dual role of both concentrator and barrier to moisture and precipitation. When air moves over mountains it undergoes expansion and compression and the

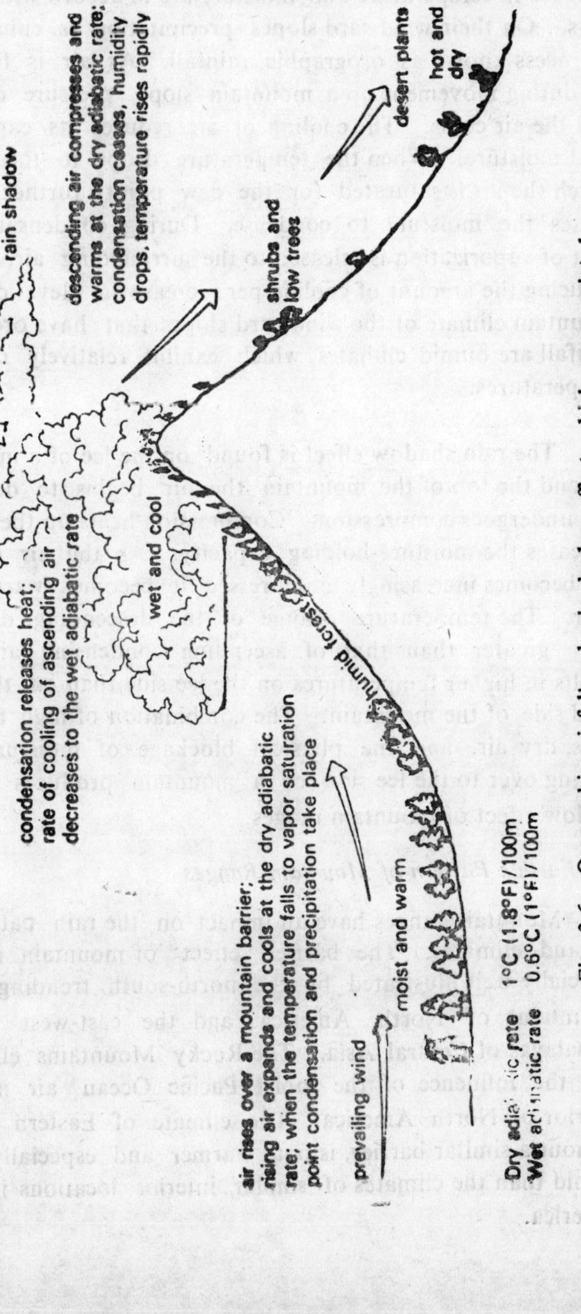

Fig. 3.6. Orographic rainfall and rain shadow effects which are associated with mountain ranges.

changes in temperature and moisture are in accord with the gas laws. On their windward slopes precipitation is enhanced by a process known as orographic rainfall. As air is forced to lift during movement up a mountain slope, pressure decreases and the air cools. The cooling of air reduces its eapacity to hold moisture. When the temperature drops to the point at which the air is saturated (or the dew point) further cooling causes the moisture to condense. During condensation the heat of vaporization is released to the surrounding air, thereby reducing the amount of cooling per increase in elevation. The mountain climate of the windward slopes that have orographic rainfall are humid climates which exhibit relatively moderate temperatures.

The rain shadow effect is found on the lee of a mountain. Beyond the top of the mountain the air begins to descended and undergoes compression. Compression heats up the air and increases the moisture-holding capacity. As the air descends and becomes increasingly compressed it becomes warmer and drier. The temperature change of the descending dry air is much greater than that of ascending condensing air. This results in higher temperatures on the lee side than on the windward side of the mountain. The combination of high temperatures, dry air, and the physical blockage of moisture from moving over to the lee side of a mountain produces the rain shadow effect of mountain ranges.

The Earth's Pattern of Mountain Ranges

Mountain ranges have an impact on the rain patterns of lowland climates. The barrier effect of mountain ranges is especially well illustrated by the north-south trending Rocky Mountains of North America and the east-west trending Himalayas of Central Asia. The Rocky Mountains effectively lock the influence of the moist Pacific Ocean air into the interior of North America. The climate of Eastern Europe, without a similar barrier, is both warmer and especially, more humid than the climates of similar interior locations in North America.

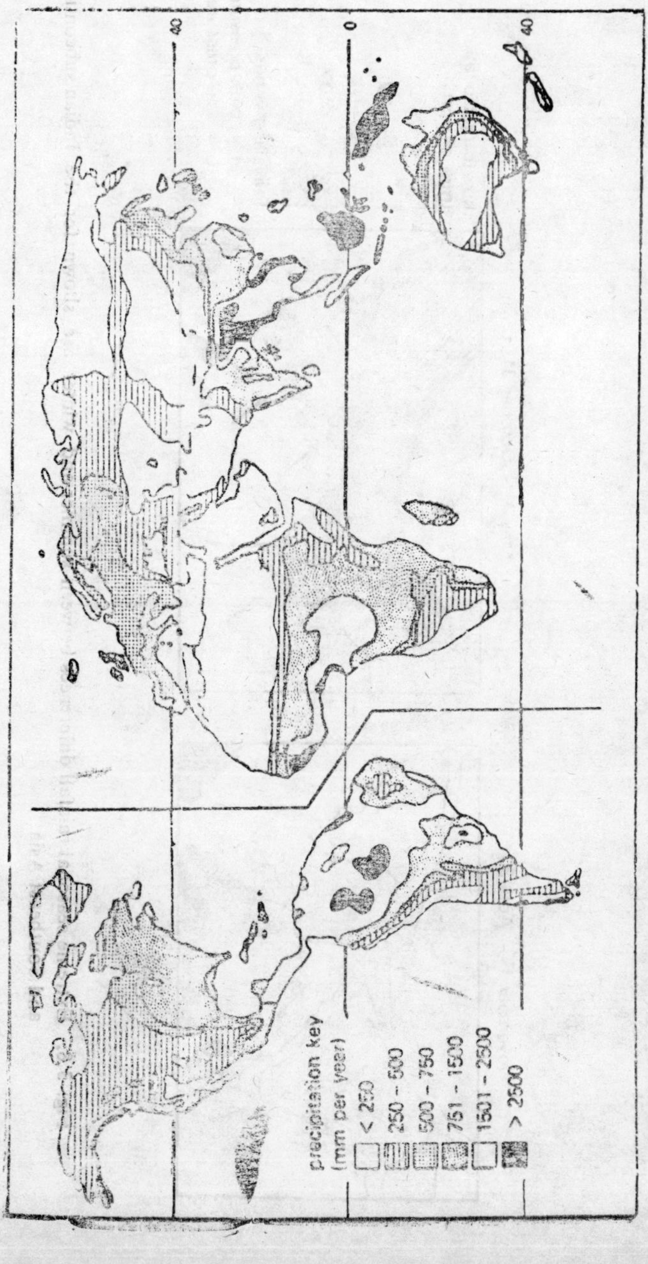

Fig. 3.7. Rainfall map of the Earth Note the influence which mountain ranges, cold water upwelling, and warm current have on precipitation in various regions.

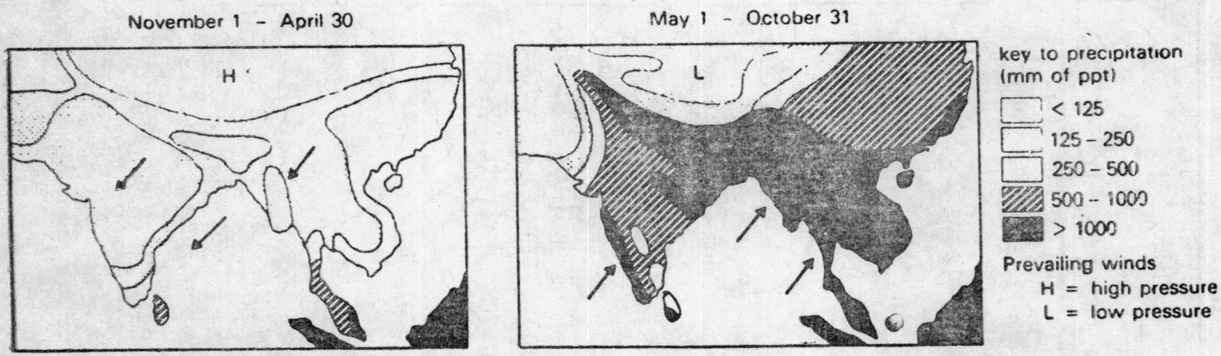

Fig. 3.8. Extreme seasonal rainfall differences between summer and winter are shown for the Indian subcontinent and Southeast Asia.

Likewise, the Himalayas act as a large barrier between the high-latitude, high-pressure air masses that build up during winter over the huge interior area of Asia north of the Himalayas and the subtropical masses to the south of the Himalayas. The climatic influence of the Himalayas is illustraed by the monsoon climate of India. The distinct change from the dry season to the wet season of the Indian subcontinent is associated with changes in position of the jet stream as it switches from the north side to the south side of the Himalayas during the cycle of the solar year. In North America, where there is no prominent east-west barrier, there is only a slight, if any, monsoon effect at a similar latitudinal location.

The role of mountains in deflecting air flow is also illustrated for tropical locations. The east and west coast climates of Africanda South America are substantially different from each other. Along the west coast of South America and the west coast of Africa, the westerlies are deflected toward the Equator by coastal mountains. The Ekman Spiral mechanism in turn operates to drive water away from the coast. The displacement of water results in upwelling of cold nutrient-rich water along the coast. The cold water strongly influences the climate of the west coast continential locations. Along the coast, warm moist air masses pass over the cold upwelling water to create huge fog banks. In the subtropics the most extreme drying occurs. The upwelling water cools air which passes over the coast and becomes compressed by the subtropical high-pressure systems. The capacity for further condensation is nil. The coastal environments of these locations are extra-ordinarily dry. In contrast, the influence of warm currents at the same latitude on the east coast of continents creates a moist climate in South America, Southeast Asia, and India. The effect is less pronounced in East Africa where highlands block the inland movement of moisture. Where east coast mountains are lacking in the Tropics (notably South America), the climate is more humid in the continental interior as a result of the fact that the warm, moist easterly air intrudes further inland.

Besides playing a role in the pattern of lowland climates by blocking and deflecting air flow, mountains are a place of important climatic processes. Mountain ranges account for 20.2% of the Earth's land surface (Barry 1981). Unlike lowlands, which tend to have a relatively homogeneous climate over a large area, mountains have the greatest climatic diversity per area of any surface feature on the Earth. At 20° N latitude near the middle of the Pacific Ocean, the island of Hawaii illustrates the influence that mountains have on climate (U.S. Department of Commerce 1967). Hawaii is an active composite shield volcano which reaches 13,796 feet (4,205 m) above sea level. From near sea level at Hilo to the Mauna Loa observatory at 11,150 feet (3,398.5 m), there is a relatively constant 6% slope. The temperature average varies at a rate of 3.3°F per 1,000 feet (0.6 °C per 100 m) of elevation. This translates into about a 0.9°F decrease in annual average temperature per mile (0.3°C/km) ascension up the slope of Mauna Loa. On the same slope the annual precipitation amounts vary by as much as 25 inches per mile (40 cm/km). A high mountain near the Equator can have the diversity of climatic situations which virtually simulate the lowland climatic conditions of the whole Earth. In mid-latitude locations the treeline has stimulated much research on conditions that limit or enhance growth. Mountains have thus been a superb site for investigating the relationships between plants and climate.

The Relationship between Climate and Vegetation

In previous chapters taxonomic relationships of plants and animals have generally been discussed. Flora and fauna differ markedly from region to region. In terms of numbers of species, the biotically richest forests of the Earth are located in the Tropics. For example, measurements of the number of tree species larger than 10 cm at breast height have been made on 1 ha plots in several places. It has been long assumed that Southeast Asia has the richest tree floras, but it is how known that the largest number ever identified, nearly 300 tree species/ha, are being recorded in the upper Amazon (Gentry 1988). At most, Southeast Asian plots sustain around 200 tree species/ha, and although comparably data are unavailable, it appears that plots in Central African rainforests match the richness of South-east Asian Tropical rainforests.

When considering global distribution, physiognomy, or external form, without reference to taxonomic relationships, has proved to be a more manageable way of grouping plants. The emphasis of this approach to biogeography is on vegetation, or the sum total of plants and their external appearance in a given region rather than on flora, or the particular species f plants of a region.

Alexander von Humboldt (1769—1859) conducted his principal work in tropical South America where he studied the relationship between plants and the environment. He described regional assemblages of vegetation which he called associations. He observed and recorded his impressions about the relationship between plant associations and elevation while at the volcano Chimborazo in the Ecuadorian Andes. He noted the similarity between the range of plant associations of a given tropical mountains and various lowland plant associations at tropical to polar latitudes. *Vegetation formations and climate.*

In view of the observations of Humboldt, it should be no surprise that plant geography became closely allied with attempts to classify climate. The early literature on the influence of climate and plant growth was expanded by the work of Alphonse de Candolle (1855). He described five plant types according to their physiological response to different climates. For plants growing in humid regions, he established four categories that were based on temperature. They were: megatherms, or plants requiring continuously high temperature, mesotherms, or plants needing moderate temperature, microtherms, or plants tolerating a seasonal cycle of warm summer and cold winters, and hekistotherms, or plants that exist beyond the treeline of the Arctic. For plants in regions of annual moisture deficit. Candolle created a special division, xerophyte. The xerophyte is a shallow-rooted plant that exploits the limited rainfall of an area by having roots extended to cover a relatively large surface area. As we shall see, Candolle's categories were extremely influential on the development of climatic classification.

During the middle of the 19th century, many botanists and naturalists collected plant specimens from all regions of the Earth. The plant specimens and information about them were brought together in several major herbariums. In 1866 a map of the distribution of some floristic features of the Earth was produced by August Grisebach. The map identified similarities in vegetation formation between widely separated floras. For example, the floras of Chile, California, the Mediterranean, and the Cape of Africa formed one group, while the floras of the Asian steppe, North American prairie, Argentine pampas and southwest African Kalahari grasslands formed another group.

A more systematic presentation of the subdivisions of the Earth's vegetation was first published by Andreas Schimper in 1898. Schimper regarded climate as the main factor influencing the distribution of vegetation. He grouped the major vegetation formations into division based on the tropical, temperate, and polar latitudinal zones. In both the tropical and temperate zones he established rainforest, forest, woodland, and grassland categories. Xerophyte or desert vegetation is included only in the tropical zone. The polar zone was the exclusive domain of tundra while the high mountain vegetation was referred to as alpine desert. Schimper differentiated between "climatic" formations, which addressed global distribution of vegetation, and "local" or "edaphic" formations, which pertain to local soil or land microclimate conditions that influence the distribution of vegetation.

The idea of using empirical information deduced from plants to produce a climate map appears in the early work of Wladimir Koppen (1870). As the basis of his classification, Koppen relied on information published by Carl Linsser (1867) which related physiological response of plants to various temperatures. In the late 1860's, Linsser published several papers, along with a map, which showed that trees dropped their leaves if the average monthly temperature dropped below approximately 18 °C (or about 64 °F).

In 1900, Wladimir Koppen published this first classification of the Earth's climate. Koppen combined the framework established by Candolle with detailed information on the physiological response of plants to temperature shown by Linsser and the information on floral similarities portrayed by Grisebach. This scheme made possible the creation of a map which reflected the temperature and precipitation distribution of the Earth, even in places where temperature and precipitation records were unavailable. The following list is a synopsis of the Koppen classification scheme and the empirical data used in its development: Candolle's and Linsser's influence is apparent throughout.

Koppen's climatic classification

(a) Megathermal climates. Tropical regimen with evergreen trees. No simultaneous leaf drop. The average monthly temperature is above 64 °F or 18 °C.

(b) Xeric climate. Desert climate with sparse vegetation Potential evaporation exceeds precipitation.

(c) Mesothermal climate. Deciduous trees, i.e., angiosperms lose leaves during the cold season. Coldest month has a temperature below 64 °F or 18 °C,but greater than 27 °F or −3 °C.

(d) Microthermal climate. Conifer tree climate or taiga. Coldest month less than 27 °F or —3 °C and warmest month above 50 °F or 10 °C.

(e) Hekistothermal climate. Tundra, Lack of trees. Warmest month below 50 °F or 10 °C.

The success of the Koppen climatic classification scheme lies in the empirical basis of the response of plants to temperature and precipitation. This has made it useful to natural scientists in many fields. Vegetation patterns fit the scheme because they were the empirical source of information for deriving the classification and establishing the boundaries. As temperature and precipitation data were gathered, they were fit into the scheme and the climatic aspects of the map became emphasized. There should be no surprise to find that the map reflects the distribution of the Earth's major vegetation formations, because from its inception, Koppen's climate map was fundamentally a vegetation map.

Consequently, maps based on the plant form or physiognomic aspect of communities are similar to Koppen's map. Those regions which have similar climatic characteristics also have vegetation that appears similar even though the actual species may be unrelated. The generalized map is based on the works of Humboldt, Candolle, Grisebach, Schimper, Linsser, and Koppen. It is not truly a vegetation map but has categories that mix climate and vegetation. With minor name changes these areas are called the biomes of the Earth. An alternative approach is to develop consistent categories for mapping vegetation alone (de Laubenfels 1975). Many of the natural patterns have been altered by human activities.

Collaboration took place between Koppen, the climatologist, and his son-in-law Alfred Wegener, Koppen and Wegener (1924) recognized that the vegetation pattern of the present Earth was organized according to principles dictated by the relationship between the Sun and the Earth, i.e., the solar climate. They argued that vegetation pattern were formed in the geologic past by the same principles as they are in the present. Accordingly, Koppen and Wegener proposed that the past position of continents could be deduced by studying physiognomic aspects of fossil plants and other evidence of past climatic conditions. For example, tropical trees do not have annual ring while middle latitude trees do. When the continents were lined up to match the tree ring features with the appropriate latitudinal zones, the configuration of Pangaea was arrived at. Wegner became a major contributor to this topic. One implication of Wegener's

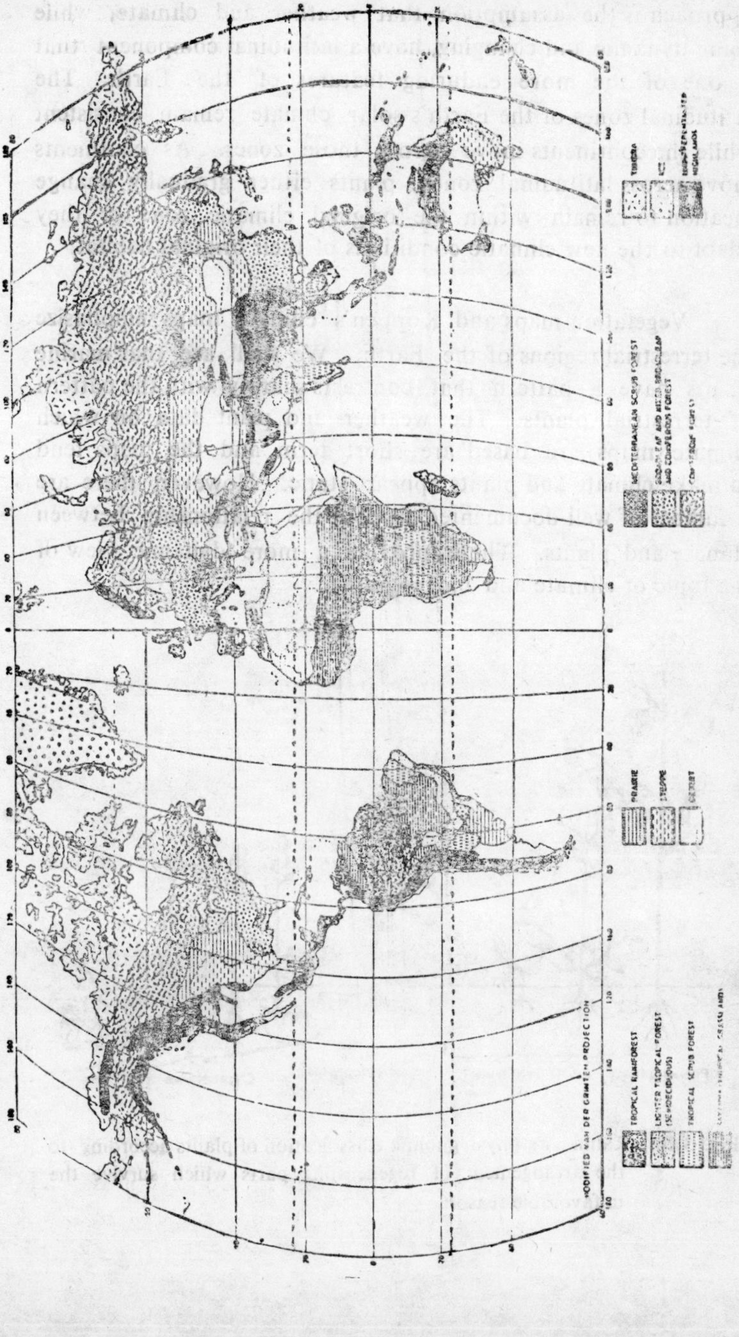

Fig. 3.9. Quasi-vegetation map of the Earth. This map closely matches Koppen's climatic map of the Earth.

approach is the assumption that weather and climate, while being dynamic and changing, have a latitudinal component that is one of the more enduring features of the Earth. The latitudinal zones of the Earth's solar climate remain consistent while the continents move across those zones. As continents move across latitudinal zones, plants either gradually change location to remain within the original climatic zone or they adapt to the new climatic conditions of their original habitat.

Vegetation maps and Koppen's climate maps emphasize the terrestrial regions of the Earth. We shall see that marine plants have a pattern that contrasts sharply with the pattern of terrestrial plants. The weather and plant data on which climate maps are based are short term and the maps tend to make climate and plants appear static. However, there are a number of well-documented cases of the relationship between climate and plants. These provide a more dynamic view of the topic of climate and biogeography.

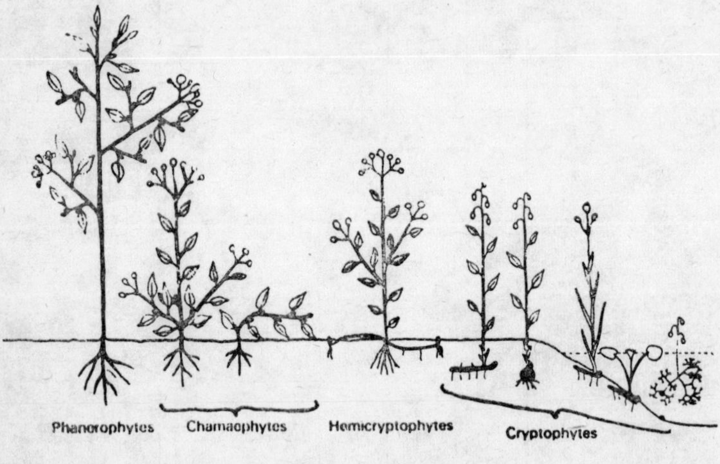

Fig. 3.10. Raunkier's physiognomic classification of plants according to the arrangement of regenerating parts which survive the unfavorable season.

Individual plants characteristic and climate

A number of physiognomic features, such as branching, leaf shape, and height, have been studied in terms of their relationship to climate. One classical physiognomic scheme was developed by Christian Raunkier (1934), a Danish botanist. Raunkier proposed that the arrangement of the regenerating parts of plants which survive the unfavorable season was a fundamental feature of plants. He assumed that there would be a relationship between the protection of plants and the severity of the climate.

Therophytes (also called ephemers or annuals) are plants which have no vegetative parts surviving from one year to the next. They survive only as seeds. Annuals are especially common in deserts of the subtropics and mid-latitude regions. Conservation of water has been carried to the ultimate limit in that no water is required to maintain the vegetative parts during the unfavorable season. As cryptophytes (also called geophytes) have surviving buds buried beneath the surface of the soil. Tuberous or bulbous herbs illustrate this form of plant. They are found in many regions. The surviving buds of hemicryptophytes are found on the soil surface. Phanerophytes have surviving buds borne on shoots that project into the air, such as on a shrub or tree.

While Raunkier assumed that the protection of surviving parts would be closely related to climate, years of examination have not produced clear patterns for plants. Several problems of interpretation have become apparent. For example, many plants such as phanerophytes belong to more than one category during their lifespan. Other factors, such as magnitude and frequency of various wind storms, drought, or ice storms provide additional insights in the relationship between plants and animals and climate.

Dynamic aspects of climate and biogeography

The frequency and magnitude of various weather events are important in the overaall dynamics of terrestrial communities, Charles Darwin (1859) suggested that ;

> the action of climate seems at first to be quite independent of the struggle for existence; but ... climate ... brings on the most severe struggle between individuals, whether of the same or of distinct species, which subsist on the same kind of food ... for example when extreme cold acts directly, it will be the least vigorous individual, or those which have got least food through the advancing winter, which will suffer most.

While the average climate provides a general measure of temperature and preciptitation on a regional basis, extreme weather events have a major influnce on vegetation patterns. Each region has a different set of meteorological conditions and different probabilities of the occurrence of particular weather phenomena. For example, the frequency of occurrence of storms such as tornadoes, hurricanes, and electrical storms, hail, ice snowfall, or frost, drought, rainfall rates, highest and lowest temperatures, range of temperatures, all to one degree or another influence the appearance and vigor of vegetation and the ability of a specific plant or animal to exist in a given location.

The frequency-magnitude model was originally described to explain the physical work done by rivers at different stages of flow (Wolman & Miller 1960). The model is both dynamic and probabilistic and it can be applied to other natural systems. Climate sets the range of conditions and vegetation adapts to those conditions. The range of conditions is extremely broad on the Earth. Climatic patterns which tie weather conditions to specific vegetative patterns have been described. In the following section the influence of zonal air-mass patterns and storm patterns and storm paterns are considered.

Air Masses

On a regional scale, a relationshp between air masses of the northern continents and plant community patterns has been described. Air masses take on the characteristics of the region over which they develop and tend to maintain those charateris-tics of the region over which they develop and tend to maintain the region over which they develop and tend to maintain those characteristics as they move into neighboring regions. The major factors which are important in the development of air masses are latitude and whether the underlying surface is marine or continental. Due to historical circumstances, some confusion over naming air masses occurred because of the fact that air masses forming over the northern Canadian and Siberian tundra were first described and named **polar air masses.** Later research revealed another air mass region over the North Pole and Arctic Ocean. Air forming over the polar regions were named **Arctic air masses,** thereby creating the present lack of correspondence between the air mass name and region of formation.

Time of the year is extremely important in the development of pressure, temperature and moisture content of air masses. Ths general position of air masses has been found to be closely related to plant communities (Bryson 1966). The average summer position of the Arctic Frontal Zone, corresponds with the southern border of the tundra and northern border of the forest. The average winter position of the Arctic Frontal Zone corresponds with the southern border of the boreal forest. Satellite mapping techniques make possible the observation of continentwide vegetation fluxes that take place during the seasonal cycle as described and employed for Africa (Tucker *et.al.*, 1985).

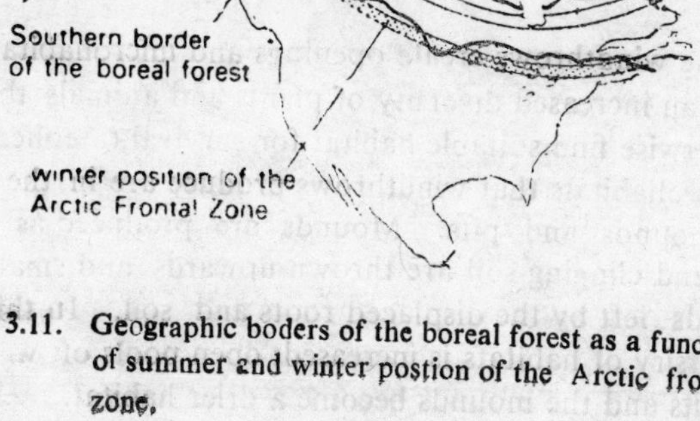

Fig. 3.11. Geographic boders of the boreal forest as a function of summer and winter postion of the Arctic frontal zone.

Storms

In the central Appalachian Mountains of the U.S.A. large floods have been found to be one of the more event for producing even age composition of forest trees within stream valleys, and floodplains. Similarly, the northern Appalachian forests have been shown to be disturbed by the occurrence of infrequent, but very largest storms such as hurricanes (Hack & Goodlett 1960). Storm have a interesting effect on plant com. munities, which seen to adapted to frequent and medium magnitude stoms however, extreme storms cause a large degree of total disruption and chaos.

Although seemingly contradictory, the disruption of plant communities by storms also has a rejuvenating effect. In the above example, the stress placed on vegetation by hurricanes creates openings for the establishment of plant species that have arrived from outside of the area as well as stimulating regrowth of the local vegetation. In this manner the disruption by hurricanes is temporary and in the long run creates a more diverse mosaic of plants than would otherwise be possible. The effect of wind on plants has also been examined (Grace 1977). Windthrows and wind maps have been studied in forest plantations in Northern Ireland and found to be extensive and related to the prevailing rather than unusual wind direction (Kennedy 1974). In fact, aerial photographs of the ground pattern of vegetation damage by severe downbursts have been the principal means for studying the dynamic processes taking place within large thunderstorm cells (Fujita 1978).

The windthrows create openings and microhabitats which support an increased diversity of plants and animals that could not otherwise find suitable habitat for survival (Stephens 1956). The microhabitats that windthrows produce are in the form of small mounds and pits. Mounds are produced as the root system and clinging soil are thrown upwards, and small pits are the voids left by the displaced roots and soil. In this manner the diversiry of habitats is increased: open pools of water form in the pits and the mounds become a drier habitat.

Another example of the relationship between climate and vegetation is what has been described as the wave generated fir forest (*Abies balsamea*) of the northeastern United States. The wave refers to an opening in the West which migrates through the forest over the number of herbs. At the leeward edge of the

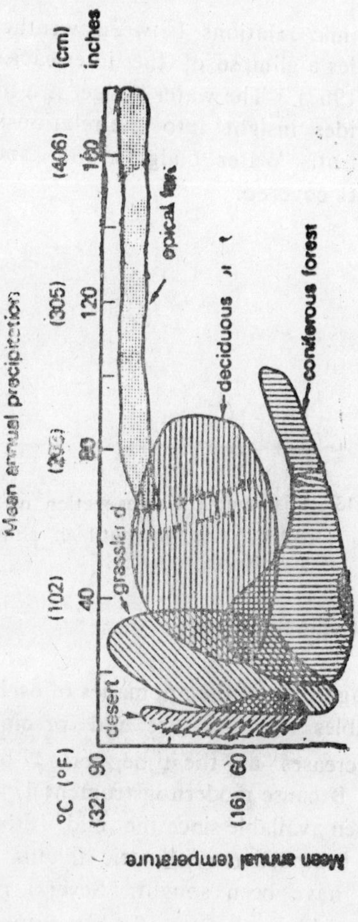

Fig. 3.12

opening trees accumulate time ice and are severely damaged as they some desiccated in the prevailing winds. As the trees at leeward edge of the opening die, they are replend by offsprings and the forest is rejuvenated. They were follow one another at intervals of about 60 years (Sprudel 1976).

The dynamic relations between weather and vegetation patterns provides a glimpse of the intericacies of an ecological system (Raup 1967). The water budget is a method of accouting that provides insight into the relationship of plants and their environment. Water budgets and land systems of the Earth the topics covered.

Fig. 3.13. Diagrammatic cross-section of a regeneration wave in a fir forest at high altitude in north east North America.

Proxy data of climatic cbange

The magnitude of the anomalies of each of the meteorological variables (temperature and precipitation and other extremes) increases as the time period being considered is lengthened. Because modern instrumental meteorological data have only been available since the mid 18th century, methods of deriving a perspective on climatic fluctuations prior to instrumental data have been sought. Several methods have been found that provide a measure of temperature and precipitation during the past. Oxygen isotope analysis of sedimentary cores is a method of obtaining temperature trends over hundreds trends over hundreds of thousands of years. Botanical evidence yields data of relatively recent climatic fluctuations. Palynology as a tool for mapping patterns of plant distribution during the glacial periods.

The method of tree-ring analysis or dendrochronology, which is sensitive to short-term fluctuation of weather variables, was pioneered by Andrew E. Douglass and first described in 1914. Douglass worked with pine trees in the semi-arid mountain ranges of the southwestern United States. Those trees are particularly sensitive to yearly fluctuations in moisture regime (Douglass 1919, 1920, 1922, Fritts 1976). Tree-ring analysis has also been applied to the eastern U.S. hardwoods as well.

Dendrochronology relies on trees growing in midlatitude climates with distinct growing and dormant seasonal cycles for a source of raw data. Each year trees form a new layer of wood during the growing season. A coring of the tree yields data both regarding age and the climatic condition of each growing season. The corings of successively older trees can be put together to create a year-to-year weather record of the region that the trees inhabit.

The use of tree rings to estimate climatic trends in the past has been accomplished by developing a meteorological model to account for various tree-ring configurations. A model useful in the southwestern U.S.A. may be stated as follows. Meteorological conditions which produce more clouds and thus cooler and wetter conditions are associated with low pressure systemes white meterological conditions which produce fewer clouds and hence hotter and drier conditions for tree growth are associated with high pressure systems. During years of frequent low pressure development, and hence cloudiness and more rain, the trees produce thicker rings, while in years with high pressure domination with cloudless skies and little rain, the trees produce thinner rings. Within a region, all trees of the same species exhibit similar patterns of broad and narrow rings. These regional patterns have made posible the tracting of regional weather systms back hundreds of years.

Tree-ring data combined with data from diaries and other historical documents have been used to reconstruct the U.S. climate back to 1601 (Fritts *et. al.* 1981). The regional climatic differences of the preceding three centuries expressed as percentage departures from the 1901-62 temperature and precipitation. means are illustrated in Figure 3.13. In general, two conclusions can be drawn: the winters for the continent have been warmer during this century than during the three preceding centuries, and during this century the North American continent has been drier on the east coast and wetter on the west coast than during the preceding three centuries.

Dendrochronology has also been developed as a tool for dating the past. A tree-ring chronology for a region is established by first dating a living tree. Ponderosa pine (*Pinus ponderosa*) has been particularly well studied because of its use in building structures by American Indians. The pattern of the earliest formed rings of a modern tree are matched to the outer rings of an older piece of wood. In the southwestern United States, structural beams of pueblos and cliff dwellings constructed by American Indians have been used to develop a chronology which extends back to 322 B.C. The oldest tree-ring chronologies have been developed for the bristlecone pine (*Pinus aristata*) and extend more than 7,000 years into the past. The wood of dated tree rings has been analyzed for various carbon has contributed toward understanding solar activity in the past and also to correct the standard carbon dating curve.

Causes of climatic fluctuations

The causes of the fluctuations of climate were considered in terms of the Milankovich model as it relates to continental glaciation during the Pleistocene. Climatologist also recognize other events that may have implications for short-term fluctuations in climate. Two events that are likely to contribute to the establishment of cooling trends include phreatic volcanism and changes in solar output.

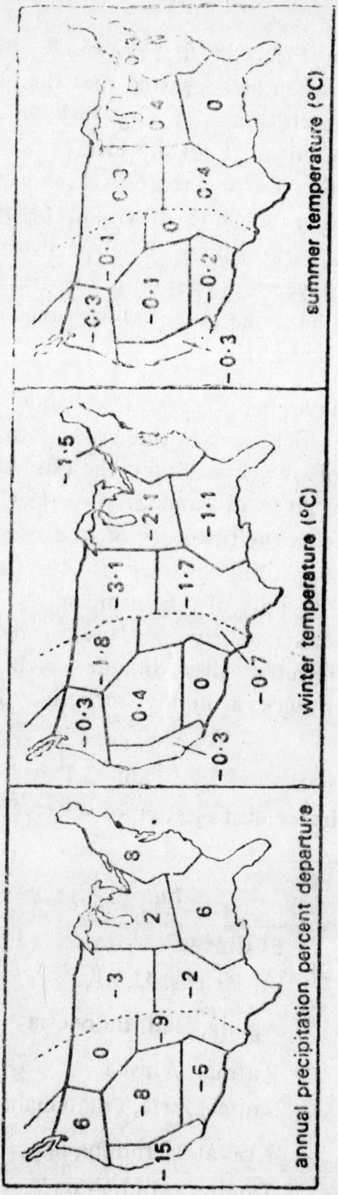

Fig. 3.14 Annual precipitation and summer and winter temperature departures.

Volcanism

Benjamin Franklin first proposed that volcanoes might affect weather when he suggested that the harsh winter of 1784 might be the result of a "dry fog" which originated with the volcanic eruption of Laki in Iceland the previous summer (Franklin 1786). Large phreatic volcanic eruptions are responsible for injecting volcanic dust and aerosols into the upper troposhere and stratosphere. The particles remain aloft from one to several years, depending on the size and altitude of injection, and have the potential for influencing global climate by reflecting sunlight.

There have been several volcanic eruptions during the past century which have injected ash into the atmosphere. A list of some of the events and the dust loads ejecteed into the atmosphere is given in Table 3.1. Most of these events had varying influences on the turbidity of the atmosphere. In addition to dust loading, sulfur dioxide loading may be significant in changing the turidity of the atmosphere. The eruption of El Chichon was particularly important in reducing incoming solar energy, It injected sulfur dioxide which Combined with water vapor to produce a highly reflective fog in the upper atmosphere.

Table 3.1

Volcanic dust ejection of selected volcanic events

Year	*Volcano*	*Dust ejected (km)*
1982	El Chichon, Mexico	27
1980	St. Helens, U.S.A.	27
1963	Agung, Bali, Indonesia	100+
1917	Katmai, Alaska	12
1902	Santa Maria, Guatemala	5
1883	Krakatoa, Indonesia	1,000+
1812	Tambora, Indonesia	10,000+

Sources: Simkin *et. al.* 1871, Mitchell 1982

The events of Table 3.1 are small compared with past volcanic eruptions, which have covered thousands of square kilometers with ash and rock deposits to depths of tens to hundreds of meters in western North America, Central America, New Zealand, and Japan. Some of the past volcanic eruption had rock volumes that ranged from hundreds to thousands of cubic kilometers (Decker & Decker 1981, Simkin *et, al.* 1981). There is evidence that in the past, periods of active volcanism accompanied glacial episodes (Bray 1977). The role of volcanism in establishing conditions leading to the accumulation of ice is not well understood.

An alternative explanation of the relationship between volcanism and climate has been proposed. Dust veils may provide for short-term global cooling (less than 10 years), but seem unlikely to account for longer-term fluctuations in the 10–100 year range. Climatic fluxes may cause stresses on the Earth's crust by loading and unloading of ice and water masses and thereby cause long-term seismic and volcanic events (Rampino *et al.* 1979).

Solar variation and tree rings

Some researchers report a connection between short-term (11 and 22 year) sunspot cycles and climatic variables, especially precipitation. However, there is much controversy over the physical connection between solar storms and the Earth's weather. The most convincing relationship between solar activity and whether is the linking of solar minima with cooling trends, and solar maxima with warming trends.

Tree-ring analysis has provided the method for estimating solar activity during the past 7,000 years. Prior to the nuclear age, radioactive carbon-14 varied directly with long-term solar storm activity. When the Sun is more active, as during intense solar storm activity, the magnetic field of the Sun shields the Earth from cosmic rays and the radiocarbon production falls. When the Sun is less active, as during sunspot minima, the magnetic shield of the Sun is reduced and the Earth receives increased cosmic radiation, which in turn increases the radiocarbon production. The short-tem solar cycles do not register in the radiocarbon fluxes. However, extended periods of changes in solar activity such as during the medieval maximum, the Sporer minimum from 1450 to 1540, and the Maunder minimum of 1645 to 1715 do show up in the tree-ring data (Eddy 1981). The maximum corresponds with a particularly warm period and the minimum corresponds with climatic cooling trends. Overall, the physical connection between solar cycles, weather, and climate is poorly understood.

El Nino and the Southern Oscillation

During a typical year the cold upwelling coastal waters of Ecuador and Peru undergo warming during December as a result of a southward flowing current of equatorial waters into the region. Local fishermen called the warm water current "El Nino" (the Christ child) because of its arrival around Christmas and the usually beneficial qualities of the warm current. It carries tropical fauna and flora that are exotic to the cold coastal waters and is pleasant.

An irregular and catastrophic form of El Nino occurs about once every 4 years, with intervals as short as 2 and as long as 10 years. These events produce massive warming of the coastal waters and disrupt the upwelling food chain to the extent that there is widespread mortality of fish and birds in the coastal waters. The term El Nino is now reserved for these disruptive events. The warming of waters off the coast of South America is now recognized as being directly connected to changes across the entire tropical Pacific and indirectly to changes in the meteorology of the whole planet (Cane 1983).

Under the normal conditions of cold water upwelling, large atmospheric pressure differences develop between the air above the cold water in the southeastern Pacific and warm equatorial waters across the Indian Ocean between Africa and northern Australia. These pressure changes are known as the Southern Oscillation. During an El Nino event the cold currents become significantly warmer resulting in less of a pressure difference between the two regions and an abatement of winds. Thus, it became clear that El Nino and the Southern Oscillation were inextricably related to each other and to pressure, wind, temperature, and rainfall within the Tropics (Rasmusson & Wallace 1983, Vaills 1986).

The El Nino event of 1982-83 was unusual and especially severe. It was anomalous because El Nino was out of phase with the Southern Oscillation (Philander 1983). In July 1982 the coastal waters were normal but by October they were 5°C (9°F) higher than normal at the Galapagos Islands. Many weather anomalies such as excessive rainfall in the coastal deserts of California and Peru, drought in Australia and Africa, and extreme cold in the interior of North America and northern Eurasia have been tied to the 1982-83 El Nino-Southern Oscillation phenomenon (Kerr 1982. Rasmusson & Wallace 1983), An opposite condition of colder than normal water in the tropical Pacific is called La Nina and causes the opposite effects. The last La Nina began in 1975 and contributed to very cold weather in 1976. A similar condition was developing in 1988. (Kerr 1988). These short-term temperature and precipitation fluctuations place extra-ordinary stress on biological communities.

Conclusions

In this chapter the Sun's role in controlliug the Earth's climate (i.e., the solar climate) and its influence on the pattern of vegetation was described. The solar climate has apparently remained within a relatively narrow range of temperature during the period of time that organisms have inhabited the Earth. The solar input outside of the Earth's atmosphere is distributed in a pattern that is directly related to latitude on the Earth. However, the surface pattern of heat and precipitation does not strictly follow a latitudinal pattern. The Earth's pressure-generated wind patterns are influenced by the atmosphere/ocean interface, Coriolis Effect. Ekman Spiral, different characteristics of land and water, and distribution of land masses and mountain ranges. As a consequence, the resulting spatial pattern of the elements of weather, such as winds, temperature, and precipitation set up complicated patterns on the Earth.

Organisms have evolved many distinctive adaptive characteristics in order to inhabit the large variation of climates on the Earth. There are many examples of similar climates being inhabited by unrelated but phosiognomically similar plants. The relationship of climate and vegetation is complicated by dynamic events. Within a specific climatic region, magnitude and frequency of extremes of weather also shape the appearance of individual plants and plant communities.

Ciimate undergoes continuous change. Fluctuations of maximum and minimum solar flare periods, volcanic eruptions, and EI Nino are major causes of decade-to-decade climatic changes. The fluctuation of climate is recorded by trees in their annual production of new wood and in pollen deposits of the species of the community of plants. Taken together, all of these observations reinforce the general concept that climate is a primary factor in shaping evolution and creating conditions for the present biogeograhic patterns on the Earth. Two major subdivisions, the marine and the terrestrial biosystsms of the Earth.

4

Freshwater Environments

Introduction

The aquatic systems of the terrestrial environment are, per area, among the most productive systems of the Earth. The purpose of this chapter is to examine the features and processes that contribute to their high productivity.

The freshwater environments include streams and lakes, as well as wetlands such as peatlands or bogs, marshes, and swamps. Aquatic features such as lakes and wetlands depend upon climatic conditions, geological events, or biological actions to develop and maintain bodies of water. These sites are thus geograhically restricted. Estuaries develop where fresh and sea water meet and are also highly localized : their location is limited mainly to the mouths of rivers. Streams are far more common, and because they are the most ubiquitous terrestrial aquatic feature on the Earth, they will be the main focus of the chapter.

The water cycle

The water cycle, or **hydrologic cycle**, refers to the exchange of water between the ocean and atmosphere. On the Earth, the oceans, which cover 71% of the surface, are the major reservoir of the water. The oceans lose water by evaporation and gain water by precipitation. This is where the process becomes intriguing. The oceans only gain about 90% of the water that is evaporated. The remainder is transferred to the land and returns to the oceans as runoff. The total amount of water involved in the cycle, about 0.04% of the Earth's water, represents all of the water that appears in the atmosphere (some of which appears as rain), the rivers, lakes, wetlands, soil moisture, and groundwater. At any one time, under present climatic conditions, 97.5% of the water on the planet occurs in the oceans.

The remaining 2.5% is divided among the following categories. About 80% is in the ice caps and glaciers and about 20% is groundwater. Lakes contain about 0.7% of the water, soil moisture holds about 0.44% of the water, and the atmosphere and streams contain about 0.05 and 0.005% respectively of the non-ocean water. Table 4.1 lists the approximate residence time, percentage, and amount of each reservoir ; it illustrates the relatively small portion of total water involved in the land part of the cycle. The land area occupied by open water also provides perspective to the water cycle. Lakes occupy about 2.5 million km^2 or 1.7% of the surface area of the continents of the Earth (Czaya 1981). They contain about 241,000 km^3 of water. The major lakes of the Earth, i.e., Lake Baikal of Asia, the Great Lakes of North America, and the Rift Valley lakes of Africa contain almost three-quarters of the fresh water. All of the remaining lakes, ponds, and wetlands share the remaining portion. Streams on the other hand occupy less than 0.1% of the area of the land (Hynes 1970). Annually, rivers, which occupy about one-seventeenth of the land area of lakes, discharge about 37,000 km^3 of water back to the ocean, or nearly one-seventh as much water as stored in lakes. Annually, streams, with only a fraction of the depth of lakes, handle about 2.6 times more volume of water per occupied area than lakes. That in essence describes the stream—a large volume of water flowing along a narrow ribbon of land.

Streams

At any one time, streams and rivers contain about 1,230 km^3, or about 3.3% of the annual volume of water which

Table 4.1

The major water reservoirs of the Earth

Reservoir	*Percent*	*Residence time*	*Amount (1,000's km³)*
Total water	100.0	—	1,382.000
Oceans	97.5	1.000's years	1.347.000
Ice	2.0	1.000'-1.000.000's years	276.000
Groundwater	0.5	10-100's years	6.908
Lakes	0.0175	variable	241
Soil moisture	0.011	months	152
Rivers	0.0001	days	1.2
Atmosphere	0.001	10's of days	12.3

returns to the seas. Stated differently, during the course of a year, the Earth's streams transport to the sea about 30 times more water than they contain at any one time. There are many possible transfers of water from streams. Some water undergoes evaporation and some seeps as groundwater into the spaces between particles of sediment. Some of the groundwater enters the root zone to become available to plants and thence undergoes evapotranspiration. The relationships between soils, plants, and the hydrologic cycle are described in greater. However, before these relationships are examined, the characteristics of streams will be considered.

Streams are located at virtually all latitudes and in all climates. Climate delivers the crucial ingredients which determine the regional characteristics of streams. Streams from under conditions where precipitation exceeds evaporation and infiltration. There are several types of streams. **Perennial** or **effluent** streams occur in humid climates where the supply of water is sufficient to flow in every season. The smallest drainage basins of the upper reaches of a large effluent river system often have **intermittent** streams. These small drainage basins characteristically have large variations in stream flow because they have too little storage of water to continuously supply the stream with water. They may flow, for example, for only a few days after each input of precipitation. **Ephemeral** streams occur in arid climates where infrequent precipitation and high evaporation leads to inadequate water for sustaining stream flow over long periods of time or along extended channel courses. The water in these streams tends to soak into the channel bed and such streams are called **influent** streams. In reality, these categories are complicated by the fact that the Earth's largest drainage basins extend across several major climatic zones.

The drainage basin

Stream channels occur in a reasonably well ordered pattern. Most streams are part of a drainage network that, connected, remove water from a drainage basin. Generally, geological conditions determine the geometry of the drainage network. Thus, the drainage network provides important clues to the geologic nature of the terrain.

For biogeographical purposes, the most important pattern of the drainage basin shows up in the longitudinal profile of a stream. Most of the majpr streams and rivers of the Earth have a slope profile that is steepest at its headwaters, and has an exponential decrease in slope toward the mouth.

The drainage basin of a large river consists of many smaller drainage basins containing smaller streams. A scheme has been devised to classify drainage basins according to their position within the hierarchy of the drainage network. The classes of streams are called **orders**. A first-order stream basin has no tributaries. It is the top of the stream profile. Its slopes tend to be the steepest and with small tributary area, it tends to have low flows. A second-order stream basin is formed when two first order stream basins join : a third-order stream basin occurs where two second-order basins join, and so on.

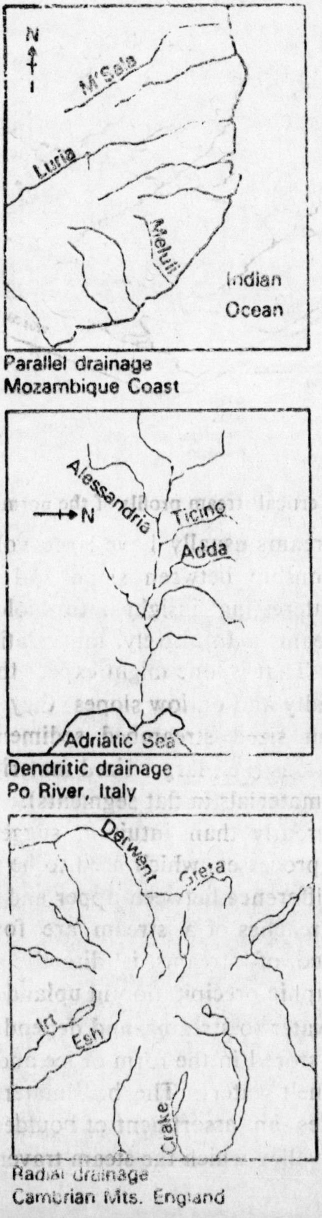

Fig. 4.1. Common drainage patterns of the Earth.

Higher order streams are associated with low slopes. Because of the number of streams that have joined to create high order

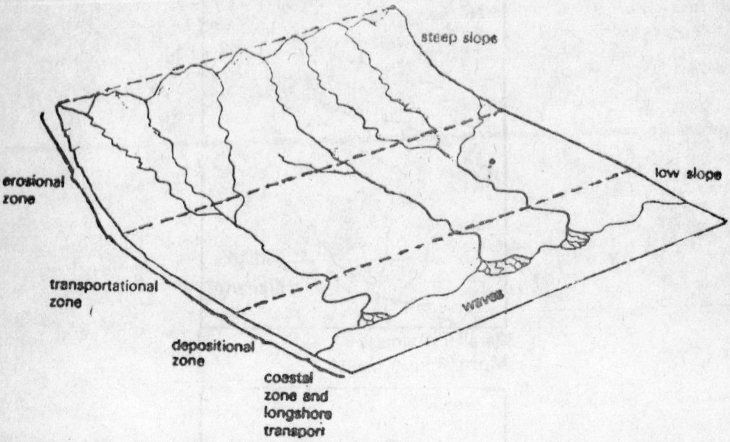

Fig. 4.2. Vertical stream profile of the normal humid river.

streams, these streams usually have large volumes of water.

The relationship between slope and stream flow rate provides an interesting insight into biologically significant properties of streams. Intuitively, the relationship may seem straighforward. That is, one might expect that on steep slopes, streams flow rapidly and on low slopes they flow slowly, and that the different sized streambed sediment at both segments matches expectations (i.e., larger sized materials in steeper segments and finer materials in flat segments). Streams, however, behave very differently than intuition suggests and there are many physical processes which need to be taken into account to describe the difference between upper and lower reaches of a stream. Upper reaches of a stream are found in highlands. The water source of streams is directly related to climate. Certainly, orographic precipitation in upland areas is important in contributing water to streams, and depending on elevation or latitude, may be stored in the form of ice and snow and appear in the steam as melt water. The bed material of the stream generally includes an assortment of boulders and other coarse materials. The valley which the steam traverses is generally a narrow, steep-sided, **V-shaped valley.** Under these conditions, a stream encounters much resistancc as it flows over the streambed. Much turbulence is apparent in the formation of various eddies and standing waves. While the water may appear to be moving swiftly, in reality, streamflow is restricted by channel resistance in its downstream progress. In the upper reaches, boulders and pebbles undergo grinding and wear as they are moved nown stream (Adams 1979).

The lower reaches of a stream, i.e., high-order stream, occur near base level, either at a lake or at sea level. Here, there may be expanses of flat land on either side of the stream which periodically become flooded, hence the name **floodplain.** The stream in these places is both wide and deep and the streambed particles consist of silts and clays. The streambed surface is thus relatively smooth, providing little resistance to streamflow which is less turbulent in appearance. In addition, the volume of the stream water in the lower reaches creates the condition whereby most of the mass of water comes into contact with only water. Only a small fraction of the volume comes into contact with the stream bottom and sides. Under these conditions, the stream velocity on low slopes is higher than it is on steep slopes in the appear reaches (Leopold *et al.* 1964). Based on a continuation of bedload characteristics, slope, amount of turbulence, and flow, the stream is naturally divided into upper and lower segments. These physical divisions have distinctive properties that are important to organisins.

Regional variation of drainage basins

dealt with steam characteristics that generally apply to all drainage basins. There are some characteristics that distinguish drainage basins from each other. In general, most of the major distinctions occur as a result of variations in climate and geology.

Climate clearly influences the characteristics of a given drainage basin. In both the Blue and White Nile headwaters, there is ample precipitation. In the headwaters, the Nile is a normal efficient stream which receives a portion of its flow from soil and groundwater. However, as the Nile flows across the wet-dry Tropics and into the desert the situation changes. The last major tributary entering the Nile is the Atbara, over 2000 km from the mouth of the Nile. From there on, the Nile becomes an influent stream, whereby it feeds groundwater reservoirs and undergoes evaporation during its entire journey across the Sahara Desert. In this manner, runoff is continuously diminished from the flow that originates in the upper reaches of the stream. A stream which passes through a desert for a major portion of its length is called an **exotic** stream. Historically, much confusion arose about the nature of streams because of the fact that the Nile appeared to come from a desert, and even more amazingly, annually flooded the land even though it came from the desert.

The major streams of the Earth listed in Table 4.2 are not exotic. They receive ample rainfall at their source and throughout most of their course, thereby gaining volume on their route to the sea. Most of the great rivers exist in humid environments. Some of the most voluminous rivers such as the Yenisei and Lena are not located in particularly humid environments. They drain an exceptionally large region, however, and they exist in a cool climate with a relatively low rate of evapotranspiration.

These drain into interior basins such as the Dead Sea. Death Valley, Caspian Sea, and Lake Eyee. With the exception of Australia's Lake Eyre, these lakes are located below sea level. They all exist in climates where, overall, evaporation exceeds precipitation. These inland basin lakes of the Earth are the recipients of stream water from surrounding drainage basins. They illustrate the point that if a stream does not receive enough precipitation, if cannot carve a path to the ocean. Many interior basins of the Earth are intriguing because they contain a biota that is relict from a time when the drainage basin was connected to a large network of streams during more humid times. These basins provide an opportunity to study evolutionary trends that occur in isolation, in addition to those presented by oceanic islands (Naiman & Soitz 1980).

Table 4.2

The major river basins of the Earth (ranked by dischange) (after McCullough 1979)

River	Length (km)	Basin area (km)	Discharge (km^3/y)	Rainfall intensity (mm/y)	Dissolved transport (tkm^2/y)	Solid transport (tkm/y)	Amount transport ($tm^2 10^6$/)y
1. Amazon	6,303	7,049,980	3,767.8	534	46.4	70.0	290.0
2. Congo	4,373	3.690.750	1,255.9	340	11.7	43.2	47.0
3. Yangtze	5,529	1,959,375	690.8	353	NA	490.0	NA
4. Mississippi—Missouri	6,215	3,221,183	556.2	173	40.0	94.0	131.0
5. Yenisei	4,991	2,597,700	550.8	212	28.0	5.1	73.0
6. Mekong	5,186	810,670	538.3	664	75.0	485.0	59.0
7. Orinoco	2,066	906,500	538.2	594	52.0	91.9	50.0
8. Parana	3,874	3, 102,820	493.3	159	20.0	10.0	56.0
9. Lena	5,853	2,424,017	475.5	196	36.0	0.3	85.0
10. Brahmaputra	1,610	934,990	475.5	509	130.0	1,370.0	75.0
11. Irrawady	4,152	431,000	443.3	1,029	NA	700.0	NA
12. Ganges	1,610	488,992	439.6	899	78.0	537.0	76.0
13. McKenzie	3,276	1,766,380	403.7	229	39.0	65.0	NA
14. Ob	5,508	3,706,290	395.5	107	20.0	6.3	50.0

(Contd.)

15. Amur	4,347	1,843,044	349.9	190	10.9	13.6	20.0
16. St. Lawrence	2,512	1,010.100	322.9	320	51.0	8.0	54.0
17. Indus	2,898	963,480	269.1	279	65.0	500.0	68.0
18. Zambezi	2,737	1,329,962	269.1	202	11.5	75.0	15.4
19. Volga	3,689	1,379,952	256.6	186	57.0	19.0	77.0
20. Niger	4,186	1,502,200	224.3	149	9.0	60.0	10.0
21. Columbia	1,955	668,220	210.8	316	52.2	42.0	34.0
22. Danube	2,861	816.990	197.4	242	75.0	84.0	60.0
23. Yukon	3,186	865,060	193.8	224	44.0	103.0	34.8
24. Fraser	1,368	219,632	112.4	512	NA	NA	NA
25. San Francisco	3,199	652,680	107.7	165	NA	NA	NA
26. Hwang-Ho (Yellow)	4,671	1,258,740	104.1	83	NA	2,150.0	NA
27. Nile	6,693	2,849,000	80.7	28	5.8	37.0	10.0
28. Nelson	2,576	1,072,260	76.2	71	27.0	NA	31.0
29. Murray-Darling	5,427	1,072,908	12.6	12	8.2	30.0	2.3

Note : NA = not available

In addition to influencing the flow volume of a stream, climate plays a related role in establishing the physical conditions that cause weathering and consequently the accumulation of various minerals in soils. Tropical soils tend to be rich in aluminum and iron oxides and poor in cations such as calcium, magnesium, sodium, and potassium. Tropical streams tend to be devoid of base minerals and are thus very soft water. During flooding the sediments carry the aluminum and iron oxides into tropical rivers. Likewise, middle latitude dry soils are enriched in a variety of base elements. The streams of these region tend to be rich in base elements and thus are hard water. During flooding, the stream sediments are enriched in base-rich elements. In humid middle latitude climates, the soils may be heavily leached and the waters arising from them poor in base elements and therefore soft. High-latitude soils are rich in silica and the streams reflect this fact.

Streams may tend to mirror soil-forming processes but generalizations do not reflect the complexity of the actual chemical characteristics of streams. Streams are strongly influenced by regional or local geologic conditions. The water quality of the stream is determined by minerals with which it comes into contact. For example, the Colorado River passes through a salt dome in its course to the Gulf of California. In doing so, the river picks up sodium chloride. In an arid climate, the sodium chloride becomes concentrated during evaporation, and this influences water quality and the tope of organisms that can exist in the river, especially during droughty years. Some rivers flow through linestone formations and pick up calcium carbonate. The dissolved carbon dioxide (CO_2) in water readily reacts with calcium carbonate to form soluble calcium bicarbonate. Calcium bicarbonate, however, stays in solution only as long as there is a certain level of free CO_2. Any condition, such as photosynthesis, which removes carbon dioxide from water also causes the precipitation of calcium carbonate. The calcium carbonate precipitate builds-up a crystalline deposit around cells of algae. These deposits of calcium carbonate are called traver-

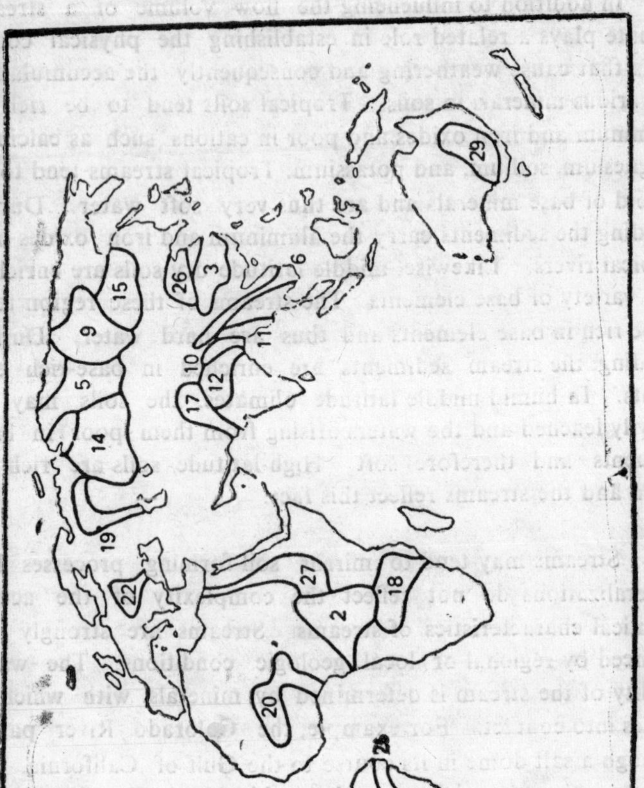

Fig. 4.3 A map of the Earth's major river basins as listed in Table 4.2.

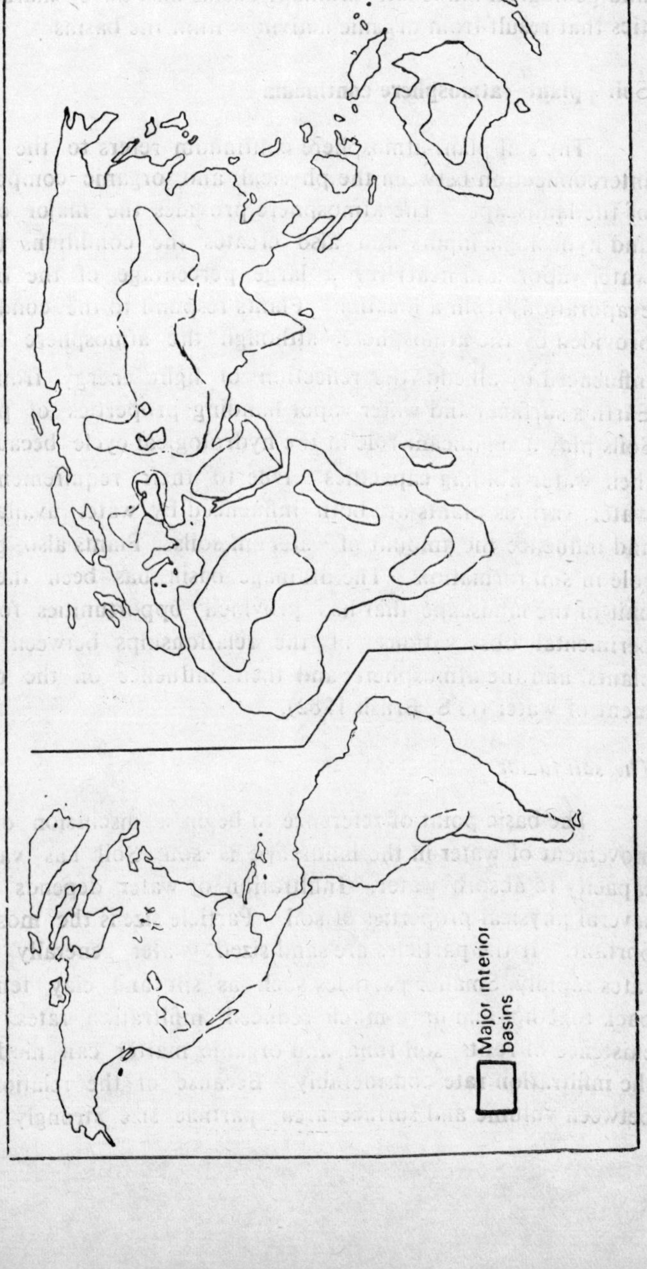

Fig. 4.4 The major interior basins of the earth.

tine (Hynes 1970). Such rivers have an unusual flora and fauna. Apart from physical conditions that arise from various regional and geological influences, drainage basins also have characteristics that result from organic activity within the basins.

Soil – plant – atmosphere continuum

The soil-plant-atmosphere continuum refers to the many interconnection between the physical and organic components of the landscape. The atmosphere provides the major energy and hydrologic inputs and also creates the conditions (wind, water vapor, and heat) for a large percentage of the output evaporation) from a location. Plants respond to the conditions provided by the atmosphere, although the atmosphere too is influenced by albedo (the reflection of light energy from the Earth's surface) and water vapor handing properties of plants. Soils play a significant role in the hydrological cycle because of their water-holding capacities. Due to their requirement for water, various plants are both influenced by water availability and influence the amount of water in soils. Plants also play a role in soil formation. The drainage basin has been the key unit of the landscape that has provided opportunities for experimental observations of the relationships between soils, plants, and the atmosphere, and their influence on the movement of water (G.S. Brush 1982).

The soil factor

The basic point of reference to begin a discussion of the movement of water in the landscape is soil. Soil has varying capacity to absorb water, Infiltration of water depends upon several physical properties of soil. Particle size is the most important. If the particles are sand sized, water generally infiltrates rapidly. Smaller particles such as silt and clay tend to pack together and have much reduced infiltration rates. The existence of roots, soil funa, and organic matter can moderate the infiltration rate considerably. Because of the relationship between volume and surface area, particle size strongly influences water-holding capacity. Larger particles such as sand may adsorb water freely and then, because of a lack of surface area, be unable to retain soil moisture. Clays generally have a higher water-holding capacity because of the large surface area of the. clay particles. But, if infiltration is poor, and of the clay particles too confined, then water-holding capacity is also reduced. Silts and loams (that is, mixtures of clay and sand-sized particles) have the largest amount of available water-holding capacity reserve for plant growth (Brady 1974).

Three different types of bonding between water and soil particles are important to plants. Gravity water is that which occurs in liquid state and fills the spaces as it permeates the soil. This water is freely available to plantsand moves through the soil under the influence of gravity. Capillary water clings to small particles and fills small spaces of the soil. It is also available to plants it slowly moves into the water table. The third type of water is called hygroscopic water. It is so tightly bound to soil particles by molecular forces that it is unavailable to plants which wilt without other water. This water does not move through the soil (Brady 1974).

Soil releases water to streams and groundwater when two conditions are met. The soil must have sufficient water, and the wa:er must be under the influence of gravity. When precipitation occurs at a rate that is equal to or greater than the rate of soil infiltration, the excess water flows under the influence of gravity as runoff toward the stream. If more water infiltrates the soils than can be held by it, the surplus flows as groundwater and reappears in springs and streams, or enters an underground aquifer.

The vegetation factor

No discussion of the water cycle or the properties of a drainage basin is complete without inclusion of the organic processes and, in particular, plant modification of the movement of water. Plants influence the operation of the drainage basin to such a large extent as to be inseparable from it. The effect of vegetation on the quality and yield of water from a drainage basin is a topic that has received considerable attention (Colman 1953).

One recent study is the Hubbard Brook Experimental Forest Study (Bormann & Likens 1979). The experimental design of the research made it possible to determine the degree of management of drainage basin commodities by the biotic community. The commodities in this case include water and the major plant nutrients both in the soils and the atmosphere. A major purpose of the research was to measure the exchanges between bedrock, soil, plants, the atmosphere, and rivers. The research was conducted on six drainage basins in the White Mountain National Forest in New Hampshire. The drainage basins rest on impervious gneiss bedrock.

The studies began with the measurement of the water budget of the six undisturbed drainage basins, from 1955 through 1968. Thrnugh the use of specialized dams designed for measuring the flow rate of streams, runoff was measured. Also, the settling ponds behind the dams were used to measur- sediment movement. Provisions were made to measure precipitation, as well as dissolved mineral load of both incoming precipitation and outgoing runoff waters. The major water budget results of Hubbard Brook Forest are as follows.

Measured inputs and outputs

Average annual precipitation =123 cm

Average annual runoff =72 cm

Difference

Average annual evapotranspiration =51 cm

From these results it can be concluded that the combination of evaporation (i.e., water loss from all wetted surfaces) and transpiration (water movement from soil and groundwater to the atmosphere via plants) accounts for 41% of total annual amount of water entering the water basin (Bormann & Likens 1970). The removal of water from the soil by plants is relatively straightforward. The amount of energy required to extract soil water is small compared to the amount of energy needed to evaporate water from the leaf. Thus, up to the wilting stage, all soil water can be assumed to be equally available.

Provided there is soil moisture available to a plant, hydrostatic pressure within plant conducting tissues reaches maximum values. Stomata are openings in the leaf that control ventilation. When light is not limiting, the stomatal apertures of the leaf open to the maximum. In this way the leaf obtains the greatest possible flow of carbon dioxide. However, the condition also allows the rapid ourflow of water vapor. Because of the limitations in the amount of carbon dioxide which a plant can obtain from the atmosphere, plants use much more water during ventilation than they actually require for physiological purposes (Penman 1970).

In the humid mid-latitude climate of New Hampshire precipitation is relatively evenly distributed throughout the year; runoff, on the other hand, is unevenly distributed. Most of the runoff occurs during March through May. Peak runoff is attributed to the spring melt of accumulated winter snow. Thus, 57% of the total runoff occurs during one quarter of the year. April alone accounts for 35% of the total when rain often falls on top of the snow. At the other extreme, during August, only 0.7% of the annual average runoff takes place. The extremely low runoff during the mid to late summer can be explained by the effect of plant growth and transpiration on soil and groundwater (Bormann & Likens 1970). In spring, as the air and soil temperatures warm up, the vegetation begins to leaf out, ending the period of winter dormancy. Evaporative losses of water increase with rising temperatures, and plants begin to draw soil and groundwater at rates which exceed replenishment by precipitation.

During the peak temperatures of summer, soil and groundwater are being withdrawn by plants to the extent that almost no water is available for runoff. During low flow, small streams undergo a diurnal oscillation. During daylight, streamflow drops s evapotranspiration increases and during darkness streamflow increases. In autumn, as temperature decreases large reductions occur in plant transpiration as plants shed their leaves and become dormant again. At this time of the year, rainfall replenshes the soil and groundwater reserves and runoff from the water increases. The ground freezes and snowfall accumulates during the winter leading to a repeat of the sequence of the annual runoff cycle.

In the Hubbard Brook study, the role of plants in modifying the movement of water was further demonstrated by deforestation of one of the six drainage basins. Following deforestation, the annual runoff of the drainage basin increased by about 30-40% over the amount of runoff of the previously forested drainage basin. The greatest difference occurred during the spring and summer months when runoff values were about 400% higher than expected from the basin in its vegetated state. The difference represents the amount of water which was normally transpired by vegetation in an undisturbed drainage basin. Unvegetated, drainage-basin runoff becomes more evenly distributed during the year because it closely reflects precipitation and becomes disconnected from the influence of plant processes (Likens *et. al.* 1970).

Table 4.3

Comparison of the chemical inputs and outputs (kg/ha) of undisturbed and disturbed drainage basins

Element	*Forested*		*Disturbed*	
	precipitation	*runoff*	*Precipitation*	*runoff*
Calcium	2.6	11.8	2.3	77.3
Sodium	1,5	6.9	1.3	18.1
Magnesium	0.7	2.9	0.6	16.2
Potassium	1.1	1.7	0.5	23.0
Nitrates	4.6	1.3	5.8	104.0

Another major thrust of the Hubbard Brook study was on the biogeochemical cycle of a forested drainage basin. Comparison of disturbed and undisturbed basins edmonstrates the degree of control that organisms exert on the landscape. The chemical quality of both precipitation (input) and runoff (output) were studied on the forested and disturbed drainage basins.

In the forested drainage basin, the ratios between the input and output of the elements calcium, sodium, magnesium, and potassium range from 0.2 to 0.6. In the case of nitrates, the atmospheric contribution is about 3.5 times larger than the amount appearing as runoff. An important discovery of the research was that the loss of cations, i.e., the nutrient ions of Table 4.3, is balanced by the input of positively charged hydrogen ions delivered by precipitation. A large proportion of the hydrogen ions are associated with sulfates and nitrates that apparently originate with industrial activities. When the water containing hydrogen ions enters the soil, the hydrogen ions replace the nutrient cations attached to the exchange sites of the soil, and the nutrients are flushed out of the soil with runoff.

Water quantity and quality variation due to disturbance can be more fully appreciated by comparing drainage-basin outputs before and after deforestation. Several key points can be made. Most striking is the fact that the disturbed drainage-basin biota loses control over the movement of elements in a local region. All of the nutrient elements were being released at a rate which was between 14 and 46 times greater than gained via precipitation. Many other changes were also observed in the stream system.

Physical qualities of the water undergo fundamental changes. In the forested basin, the streams are in deep shade during summer. As a result, there is a very small daily fluctu ation of temperature of the stream water. Within a disturbed drainage basin, direct sunlight warms the soil during daytine and radiation losses during night cool the runoff. The overall change is warmer runoff which undergoes relatively large diurnal temperature fluctuation. It was also observed that the release of nutrients into the stream in large quantities (or eutrophication) caused severe deterioration of stream water quality. The combination of large quantities of nutrients and warm water triggered the rapid growth of algae in the stream. The algae in turn died and sank to the bottom and the bacay processes consumed large quantities of oxygen. The stream became anoxic and unable to support higher forms of aquatic life. In conjuction with entrophication there was increased erosion. In the deforested drainage basin, without a source of leaves the layer of litter was lost from the land surface and stream bottom. These leaf-litter substances extremely important to the stream because they form a biological barrier to erosion. Leaf-litter loss, coupled with increased runoff, caused a significant increase runoff, caused a significant increase in stream channel erosion. In sum, the drainage basin functions as a whole system in which plants and their litter play a critical role in tying the parts of the system together (Likens *et al.* 1970).

Experimental work in several locations has revealed that not only is the drainage basin influenced by plants but that the amount of streamflow is a function of the species of plant present in the basin. One long-term study has been conducted in the southern Appalachian Mountains of southwestern West Virginia (Swank & Douglass 1974). The naturally occurring deciduous oak-hickory forest was cut and replaced with a stand of white pine (*Pinus stobus*). The study demonstrated the physical differences between basins vegetated with different species of plants. The soils within a basin vegetated with conifers tended to be much drier than those vegetated with hardwoods. Conifers photosynthesize and evapotranspire much earlier in the spring than do broadleaf deciduous trees. One measure of leaf surface is the leaf area index, which is a measure of the area of leaf surface compared with the area of ground the plant is occupying. During the dormant season, conifers have a much larger leaf area index than broadleaf deciduous trees (9.9 compared to less than 1) and thus intercept and subsequently evaporate much more rainfall. Overall, the evaporation and transpiration by conifers results in a streamflow reduction of 20% compared to drainage basins composed of the broadleaf deciduous forest

(Swank & Douglass 1974). The reduced scil moisture resulting from higher transpiration rates among conifers also limits the growth of ground cover. The reduction in ground cover resulted in changes which influenced flooding cycles and water quality. During heavy precipitation, runoff of the sparsely vegetated surfaces caused an increase in soil erosion. Stream channels were deepened and shopes became steeper as an adjustment to the influence of a conifer forest. Also, the natural acidity of pine needles resulted in the leaching of cations and hence, water contained greater quantities of various elements. Type of vegetation controls basic characteristics of soil moisture, ground cover, and chemical regime, which in turn influence drainage-basin characteristics such as water quality, sedimentation transport, channel depth, and angle of slope.

Vegetation cover can influence local water availability by acting as a cloud droplet impaction network which deposits water and aerosols. Examples of this phenomenon are reported from several elevations of the northern Appalachans (Lovett *et al.* 1982) and Lomas de Lachay in coastal Peru (Holdridge 1967). In the subalpine forests of the northern Appalachians, inputs from cloud droplets account for about 46% of the water and from 150-430% more chemical inputs than bulk precipitation. In the Peruvian forest, precipitation recorded under the treecrown yielded 171% more water than was recorded in the open. The mechanism that causes this increase in precipitation is the condensation of moisture or deposition of cloud droplets on the aerial portion of the tree. The leaves act as large nets that capture droplets of moisture which do not otherwise have a possibility of coalescence to form precipitation. This process attests to the capacity of the tree canopy to influence and enhance its own microclimate. When all aspects of the relationship between plants, albedo, precipitation, and soil water are taken into consideration it becomes apparent that vegetation plays a major role in modifying the Earth's climate (Shukla & Mintz 1982, Lockwood 198).

Soil-moisture balance

The above discussion of the relationships between precipitation, vegetative cover, soil moisture, evaporation, transpiration, and runoff provides reason to consider alternative methods of classifying climate beyond the simple use of temperature and precipitation C.W. Thornthwaite (1948) developed a method for classifying climates on the basis of energy and moisture criteria. Thornthwaite's soil water balance analysis uses month by-month values of precipitation, potential evaporation, actual evaporation, and soil moisture surpluses or deficiencies to determine the water available for plant transpiration and growth at a given locality.

Potential evaporation (PE) is the amount of soil moisture which could be used by the vegetation if there was no limit to the supply of water. The principal determinant of *PE* is solar radiation. *PE* is usually expressed as the depth of liquid water that could be transpired per month. The value changes with seasons. When soil moisture is depleted and only hygroscopic water remains, the estimated actual rate of evaporation (AE) will fall below the potential rate of evaporation. By using these two estimated moisture parameters. Thornthwaite was able to chart the local deficit or surplus of soil moisture. Several problems have been observed in the calculation of water balances in various places (Major 1977, Schulze & McGee 1978). Calibrations have been developed to enhance the reliability of the Thornthwaite method in localities where they need adjustment (Shelton 1978).

One particularly useful meas ure for evalutaing the long-term availability of water resources is the moisture index (Thornthwaite & Mather 1955). The moisture index is calculated in two ways. It is the ratio of the difference between water surplus and deficit and the potent ial evaporation 100[(*SD-PE*]; and it is the ratio of the difference between precipitation and potential evaporation and the potential evapotranspiration 100[*P*/*PE*)-1]. In general, when the potential ev aporation of stations bearing particular natural vegetation types is plotted against the moisture index of the same stations, different vegetation associations become distinguishable from one another (Mather & Yoshioka 1968). A forest requires an index of about

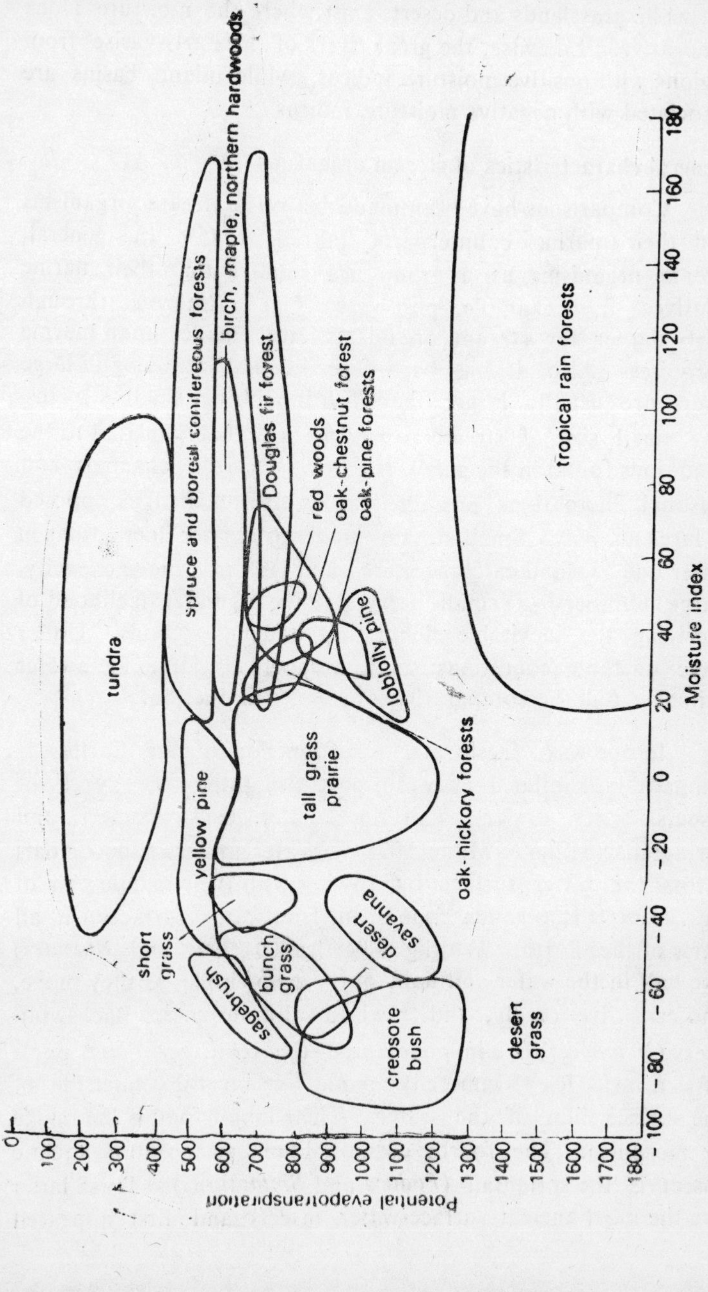

Fig. 4.5 Natural communities a function of potential evapotranspiration and moisture index.

20, while grasslands and deserts exist where the moisture index is negative. Likewise, the great rivers of the Earth arise from regions with positive moisture indices, while inland basins are associated with negative moisture indices.

General characteristics of stream organisms

Comparisons have been made between stream organisms and their marine counterparts (Patrick 1975). In general, stream organisms as a group are smaller than their marine relatives. For example, porpoises of the Amazon, through fairly larger far stream organisms, are smaller than marine porpoises. Also, it has been observed that fish living in large rivers are usually larger than fish inhabiting small streams. The small size of stream organisms is probably related to the conditions found in the small streams. Oarrow channels and seasonal fluctuations promote the survival of small as opposed to large animals. Small streams undergo greater fluctuations in water due to small catchment area and lack of storage capacity. Large numbers of small animals have a greater likelihood of surviving the inevitable fluctuation of small streams. Under these unstable conditions small numbers of large freshwater animals would be comparatively prone to extinction.

In one way, freshwater environments of the Earth ear. remarkably slmilar. They support the same four types of insects : water striders, whirlinig beetles, backswimmers, and springtails (Milne & Milne 1978). Water striders glide or dart across the water surface by rowing with their middle pair of legs. *Gerris* is a genus found on freshwater surfaces in all parts of the Earth. Whirligig beetles (*Gyrinus* and *Dincutes*) live half in the water and half out, set up ripples as they move, and can dive or fly, and crawl on solid surfaces. Backswimmers (*Notonecta*) live in ponds and the relatively quiet pools of streams. They swim belly-up and live on the underside of the surface film of the water. Their major food is the midge or mosquito. The fourth cosmopolitan surface-film aquatic insect is the springtail (*Podura* and *Sminthurus*). These latter are the most ancient surface-water insects and first appeared about 375 Ma ago. They alone inhabited the surface film until 250 Ma ago. Springtails feed on plant spores that settle onto the water surface.

Stream organisms generally have short life cycles compared with deep-lake or marine organisms. Many species have life cycles that last only one day. Since life cycles are short, the reproductive rate tends to be very high.

Stream communities tend to have long food chains that include four or five transfers of energy. Algae are important both as a food surface and as a substrate for the growth of other epiphytic food sources. Rooted plants are not as important as a food source as they are as a substrate or shelter for other organisms. They thus serve to increase habitat diversity (Patrick 1975).

Streams commonly contain very large numbers of species. Many species have developed specialized organs or appendages. for surviving in currents (Vogel 1981). Hard-water streams tend to support fewer species of algae, large number of protozoa and insects, and slightly fewer species of fish than soft-water streams (Patrick 1975). Along a single river, the number of species in common during a single collection time between different collection sites tends to be higher than the number of species in common found at a single site during different collection times (Patrick 1975) Overall, there is a high species turnover in time and space. Examination of the physical characteristics that influence biotic qualities of streams provides insight into the biogeographic *et al* atterns of the steam habitat,

Photosynthesis and respiration ratio streams

The traditional way of considering a community of organism is to regard them as a locally self-sustaining entity to to which radiant energy is supplied. The energy passes through the various trophic levels, that is, the primary producer, primary consumer, secondary consumer levels, and so on, and the decay chain breaks down the organic materials so that the constituent elements can be subsequently returned to the organisms to be recycled. This biogcochemical model may be a resonable conceptual framework for many terrestrial and even lake commodities, but it is inappropriate for the stream system. In the stream, everything released into solution moves down stream and cannot be locally recycled (Hynes 1973). A generalized approach that succeeds conveying the dynamic features of the stream is based on the longitudinal profile of the stream. There are physical features of the stream which change from the headwaters to the mouth that influence the ratio of photosynthesis and respiration. Some stream segments have almost no photosynthesis and thus a ratio of photosynthesis and respiration (P/R ratio) less then 1. The organic matter, usually leaves or needles, which supports the growth organisms here must be imported into the stream segment. These segments are principally the domain of heterotrophic organism. Other segments of the stream may have more photosynthesis than respiration and thus a P/R ratio greater than 1, and large numbers of autotrophs (Cummins 1975). The physical and biological properties of the upper and lower segments of a large stream are sufficiently different that the two parts of the stream have been subdivided into the rhythron and potamon.

Rhythron

The turbulent flowing stream in the elevated area, i.e., headwaters and mountain streams, of the drainage basin is known as the rhythron (also called rhithron), literally flow plus elevation. The rhythron is the coldest part of the stream from the source to the points where the monthly mean temperature rises to about 68-75 °F (20-24 °C), where water flow is turbulent, oxygen concentration is always high, and the streambed is composed of rocks, stones, or gravel. The headwater region of a stream is typically densely vegetated and, because of the narrowness of the stream, dimly lit. The food chain of the headwater consists of inputs of organic matter from the vegetation via leaf fall. Because the stream valley is typically steep sided in the headwaters. the quantity of leaves per cubic meter of stream per year far exceeds the quantities in lower reaches of the stream. The nutrient values vary with the type of vegetation. Species of *Alnus* (alder) have root nodules and a symbiotic relationship with bacteria similar to legumes. Thus *Alnus* leaves contain about four times more nitrogen than other deciduous species. Conifers are relatively poor nutrient suppliers to streams (Hynes 1970). The rhythron supports large

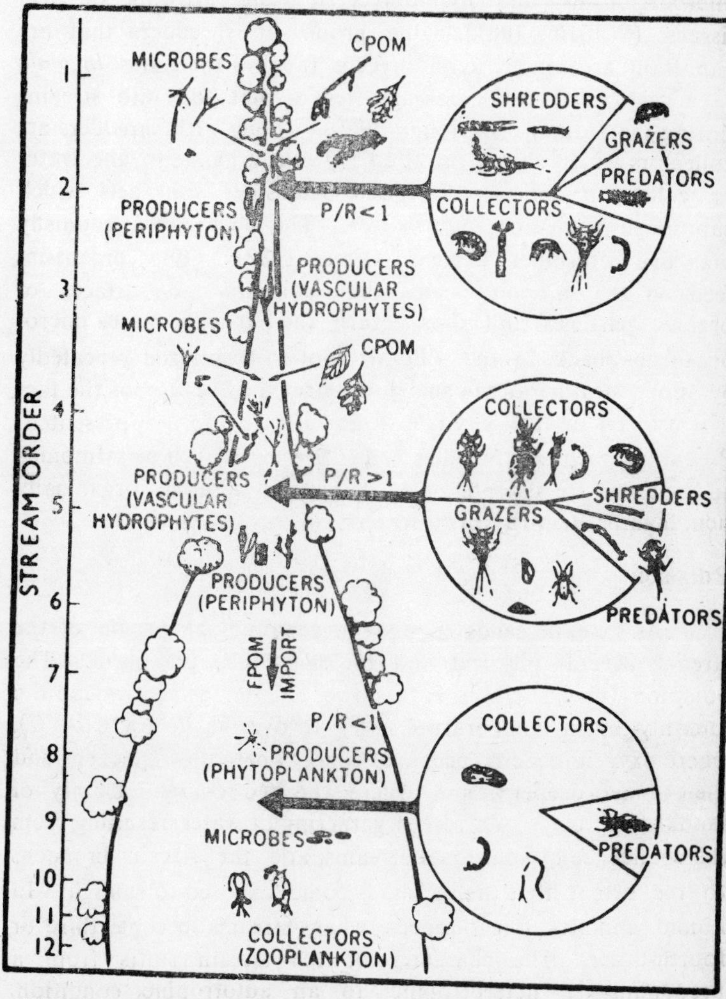

Fig. 4.6. The stream organism as a function of the stream order and position in the vertical stream profile.

numbers of benthic (bottom-dwelling) organisms specialized in shredding organic matter and another group specialized in capturing coarse particulate organic matter (CPOMM). The CPOM in turn is further disgested by these organisms and becomes fine particulate organic matter (FPOM) of the water and sediments. The shredders obtain their food source from microorganisms' and hyphomycetete fungi growing on leaf tissues (Williams 1981). One group of shredders that are mportant are aquatic hyphomycete fungi of the genus *Ingoldia* (= *Cyoerffyella*) which colonize ieaves that fall into streams during the autumn (Barloher 1982). Along with shredders are collectors which either trap the FPOM as it passes in the water or collect it from sediments. Caddisflies spin nets which capture suspended organic matter. The most ancient caddisfly taxa live in turbulent current (Alstead 1982). Other organisms feed on the microorganisms that colonize the surfaces of organic particles and thus return the particles, minus micro-organisms back to the FPOM pool to be utilized repeatedly by other benthic organisms down stream. Because of the fact that primary productivity is low, grazers are poorly represented. Predators such as stoneflies and fish species such as salmonids and sculpins are also present in these well aerated, organically rich, high-pr otein rhythron waters.

Potamon

As onc descends along the longitudinal profile of the stream, several physical and biological changes occur. The potamon (means river) is the lower stream segment where the monthly mean temperature rises to over 68-75 °F (20-24 °C), where oxygen deficits occur at times, where flow quickens and the channel deepens, and where the bed consists mainly of sand, silt, or clay. There is a gathering of water resulting from the confluence of additional streams, and the stream broadens to the extent that the middle becomes exposed to sunlight. In humid climates this happens when streams become third or fourth-order. The character of the stream shifts from a predominantly heterotrophic to an autotrophic condition. Planktonic forms are common in the surface waters, and in the benthic community, grazerf and collectors predominate. Shredders become less common. The caddisflies or more modern taxa have evolved to live in the laminar currents which have graeater concentrations of particulate food (Alstead 1982). Predators include both invertebrates and vertebrates. Collector species dominate the river benthos as a result of the inflow of large amounts of FPOM produced up stream, and locally produced FPOM from planktonic forms. Overall, the P/R ratio becomes less than 1 because high turbidity blocks sunlight. However, high turbidity may be a modern problem because many early accounts of streams along the east coast of North America indicated clarity where now there is turbidity (Trimble 1974).

The transition zone between the cooler, turbulent, shallow, and aerated rhythron and the warmer, deeper. smooth-flowing potamon generally has the largest number of species because of overlap in the ranges from both segments of the stream. The total number of fish tends to decrease while the weight per unit fish tends to increase in the downstream reaches of the stream. The weight per unit area remains more or less constant in the stream (Hynes 1970). North American rivers hava many more species of fish than do north west European rivers, though there are many similarities between the two faunas. Glaciation had an unequal influence on the faunas of the two areas. In North America, as the climate changed, the river fauna was able to retreat to warmer climates in the south. In northwestern Europe, species were caught between the ice sheet building in the north and the glaciers forming in the Alps, resulting in the extermination of many species (Hynes 1970).

The above discussion is based on research conducted primarily in humid mid-latitude climates. There are many factors which influence the productivity of a stream. For example, the seasonality of precipitation and streamflow changes with such factors as altitude, latitude, continental position, that is, all of the variables that influence climate. These same factors, because of their role in productivity, also influence the amount of organic matter available in a stream, In addition, the composition of vegetation, whether coniferous or broadleaf deciduous forests (and even the species such as alder or oak), also affects the amount or organic matter in a stream. This brings the discussion back to the water balance idea discussed earlier in the chapter. A comprehensive view of the complexity of the stream biosystem requires combining the principal factors which govern the operation of the regional water balance with balance with factors that inflence litter production and organic matter availability in the stream. The average timing of streamflow, litterfall, and peak particulate organic matter transport into streams.

The influence of latitude is equally profound. Middle-latitude streams include rhythronic and potamonic zones and support both cold and warm water adapted species. Potamonic species are capable of surviving—but not thriving—in rhythronic waters. Rhythronic animals cannot survive in the warm, ovygen-depleted waters of the potamon (Willians 1981). In high latitude environments, the streams are typically rhythronic. supporting species adapted to cold, well-aerated water throughout the entire length. In tundra freshwater systems, organic matter has accumulated in the form of terrestrial peat, and is transferred to higher organisms via aquatic insect larvae (Schell 1983).

In lower latitude streams, the rhythron may be very short or absent, with the potamon dominating the river. Large tropical lowland rivers have distinctly hlgh fish species diversity and the fish are characterized by having many adaptations to low oxygen supply (Welcome 1979). The Amazon undergoes seasonal fluctuations of as much as 20 m (Gentry A Lopez-Parodi 1980) and floods large areas during high water stages. The total efflux of organic carbon for the Amazon has provided a basis for estimating the total organic carbon input to the oceads from the Earth's rivers. The Amazon contributes about 10^{14} g of carbon and the total for the Earth's rivers must exceed 10^{15} g (Richey *ot al.* 1980). One source of organic carbon is from swamp hardwood forests and floodplains. This organic

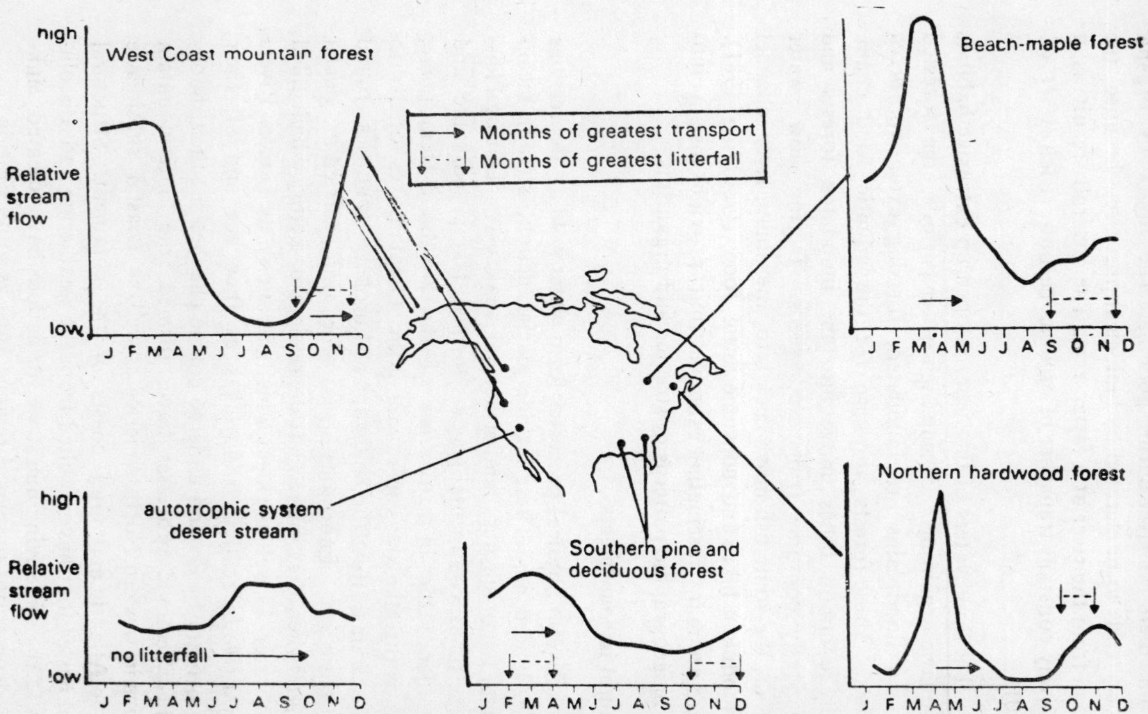

Fig. 4.7. Streamflow, litterfall, and peak particulate organic matter as a function of geographic location.

carbon enters the river during flooding. The influx of organic carbon tends to moderated by biological processes because the carbon from the river at a faster rate. The overall result is a relatively constant transport of organic carbon (Richey *et al.* 1980).

The floodwaters of the Amazon open up extensive habitats to aquatic organisms. During high water periods, the forested lowland floodplains are inundated, forming extensive *varzea* forests. These forests are base for the aquatic food chain. Many common fishes move into the inundated forests and voraciously consume fruits and seeds. During low water periods, the some fish move back into the main river channel or floodplain lakes and consume little food (Goulding 1981). There are many fascinating examples that provide insight into the biogeographic patterns of streams and organisms.

Natural impoundments

Several physical processes form water-holding depressions in the landscape. These include glacial action, stream meandering, tectonic forces, and structural weaknesses. During Pleistocene times, continental glaciers formed lake-pitted kame and kettle landscapes in many areas of the Northern Hemisphere, such as the lake areas surrounding thc Great Lakes of North America or dotting the landscape in European countries around the Baltic Sea. Enormous blocks of ice were burried in glacial till and when they melted left depressions or kettles which filled with water, as well as ridges which are known as kames. Rivers play a role by reating oxbow lakes as they cut and fill their sedimentary banks and undergo course changes in theii fiood-plains. Tectonic processes have formed the intermountain basins in western North America as well as coastal subsidence in the Atlantic and Gulf coasts of the continent. Structural weaknesses that have resulted from the weathering and dissolution of the underlying limestone formations have created sink-hole depressions in several regions, such as south Florida.

Rift lakes

The rift valley lakes of East Africa are spectacular for

their size and for the diversity of their fish fauna (Fryer and lies 1972, Beadle 1981). Lake Malawi is the southernmost rift valley lake ; it is 620 km long, 80 km wide and has a single large basin which is 750 m deep. Lake Tanganyika is even larger and deeper and has three basins, but has a smalle diversity of fish.

Lake Malawi has more species of fish than any other lake in the world. About 300 species of fish are currently known but many new species are being described, and it is estimated that at least 1,000 species probably exist in the lake (Ribbink *et al.* 1983). The most fascinating biological feature about Lake Malawi is that nearly 95% of the fish are members of a single family, Cichidae.

Cichlids are a large family of freshwater fish and are found in Southern Africa, Sri Lanka, India, South Africa, and one species is found in Taxes. Their distribuion (with the exception of the member found in (Texas) suggests a long history which extends back to a time when the continents of Gondwana were connected. All but four of the cichlids found in Lake Malawi are endemic to the lake. All species except one are mouth brooders. The exception is the red-belied bream. *Tilapia rendalli*) which digs and broods in tunnels.

The density of cichlid species is truly amazing. On Domwe and Thumba Islands near Cape Maclear, in Lake Malawi National Park there are rocky are as where densities of cichlids 15 to 20 different species of exists per 10 square meters (Reinthall & Lewis 1986). The fish have adopted a large number of specialized feeding behaviours. For example, algae and diatoms grow in the water column, on rocksand sand ; different species have specialized their feeding behavior to live in each of these conditions. In addition, some cichlids feed on the scales and fins of other fish. Lake Malawi is anoxic at about 230 m, and bacteria are the dominant organism in anoxic waters. In addition to nutrients sinking to depth and being lost, another way in which there is a loss of nutrients from the lake is through the hatching of midges (*Chaoborus* and *Chironomus*). So many midges hover in the air at one time that they form dark clouds above the surface of the lake. When the midges hatch and below onto the land they carry valuable nutrients from the lake,

As in the case of the finches and turtles which Darwin found in the Galapagos Islands the many related fish species existing within Lake Malawi raises questions regarding the development of new species.

Wetland communities

One common successional hypothesis treats open bodies of water as ephemeral features of the landscape which over time fill and disappear. Starting with an open body of water, there is a submergent stage (submergent water plants become dominant), an emergent stage (shallow water herbaceous plants dominate), and then "climaxes" with a closed stage of either a wet prairie or a wet forest. There are situations, such as within the floodplain, where this sequence may be appropriate. However, because of the qualities of the organic materials that have filled the depression, organisms may continue to play a role in depression, organisms may continue to play a role in

Table 4.4

Comparison of physical and chemical characteristics of peat and mineral soils (values in per cent based on dry matter, modified from Brady 1974)

Source	*Nature*	*Organic matter*	*N*	P_3O_5	K_2O	*CaO*
Michigan	Low lime	77.5	2.10	0.26		0.16
Michigan	High lime	85.1	2.08	0.25		6.80
Germany	Low lime	97.0	1.20	0,20	0.35	0.35
Germany	High lime	90.0	2.50	0.25	0.10	4.00
Austria	Low lime	93.3	1.40	0,10	0.06	0.45

(*Contd.*)

Austria	High lime	83.8	2.10	0.18	0.13	2.38
New York	Low lime	94.2	1.26	0.15	0.10	0.60
New York	High lime	81.5	2.56	0.19	0.21	6.51
Minnesota	Low organic	59.7	2.35	0.36	0.17	2.52
Minnesota	High organic	94.0	1.70	0.16	0.04	0.31
Florida	Sawgrass peat	87.1	2.79	0,41	0.04	5.20
Canada	Peat soil	74.3	2.19	0.20	0.16	--
Washing-ton	Woody sedge peat	89.2	3.52	0.43	0.09	1.29
Mineral soil (typical)		4.0	0.15	0.04	1.70	0.40

maintenance of the body of water. There are a number of wetland communities of plants and animals which thrive within and are an inextricable component of the drainage system. A few of these communities, the bog, marsh, and swamp, are briefly described below.

Bogs

Two broad groups of soils are commonly recognized : mineral and organic soils. Organic soils are composed of large accumulations of mosses or sedges, and are commonly referred to as peat. They have the highest orgonic matter content of any soil, see Table 4.4. One of the outstanding physical properties of the materials is their high water-holding copacity. Mineral soils hold about 20% to 4% of their weight as water while undecayed moss or sedge peat can hold 12 to 20 times their dry weight as water. These materials are veritable sponges in the landscape.

Organic deposits of peat occur in many places on the Earth. They are located in structural basins, such as the Everglades of Florida, which has sedge-derived peat, but the

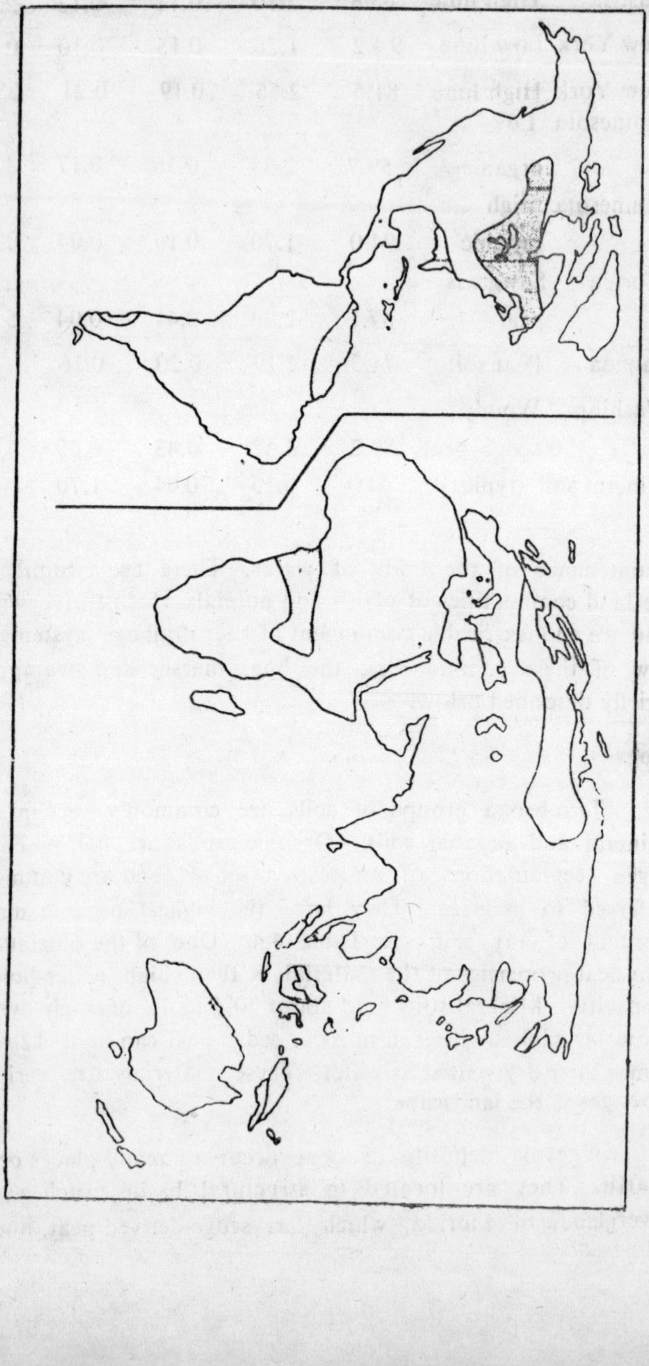

Fig. 4.8. Location of the major peat soils of the Earth.

major locations are in glaciated landscapes. Extensive deposits occur in northwestern Eurasia and North America.

Sphagnum moss composes the peat of the bogs of large areas of North America and Eurasia. The water in these bogs is cold, anaerobic, high in dissolved organic matter, and low in minerals. The moss is essentially sterile, containing neither bacteria nor the organisms of decay that are usually associated with soil. The large bogs exert much influence on the hydroxy of the surrounding region. They may cause either an increase or decrease on the annual spring flood peak, depending on how the bog is tied into the groundwater system. In general, the bogs reduce storm peaks significantly and augment streamflow during dry periods. Thus, rivers that are supplied by the bogs tend to have a more constant flow than other streams. Bog Communities thus exert a moderating influence on the stream system.

Marshes and Swamps

Marshes and swamps develop in moist climates, generally in low lying depressions, and play an important role in channelling the runoff of a stream system. If the depression is shallow and poorly drained, and supports plant that emerge above the water surface, it qualifies as marsh or swamp. When water enters the depression, the velocity decreases rapidly, and the organic matter, sedimentary particles, and minerals that have been carried by the stream flocculate and accumulate within a relatively small area. These small areas are the recipients of the nutrients and organic wealth of the watershed and are exceptionally rich places on the Earth.

Marshes and swamps are distinguished from each other by their vegetation. Vegetation varies with climate. The seasonal, continental mid-latitude climates support cattails and reeds that make up the marsh. warmer climates support the growth of trees such as cypress which make up the swamp vegetation.

Emergent marshes are extremely productive environments. They annually produce between 450 ane 2,320 g/m² (dry weight) of above-ground material and compare favorably with saltwater marshes in their productivity (Good *et al.* 1981). In order of their productivity are cattails (*Tapha*). reed canarygrass (*Phalaris arundinacea*) reeds (*Phragmites communis*), sedges (*Carex*), and bulrush (*Scirpus*) (Pratt and Andrews 1981). Marshes can undergo enormous fluctuation of productivity from year to year. Prairie marshes have been described which vary by as much as 18-fold over a period of 5—20 years (Good *et al.* 1978). Marsh variability is related to two major factors, First, weather fluctuation leads to changes in water depth that strongly influence the plants which can be supported within a body of water. During wet years, high water may flood out emergent plants, while dry years may result in the filling in of a marsh with emergent plants. Severe drought may even lead to the drying up of a prairie pothole, whence the unprotected bottom may become depened by wind erosion. In North America an animal plays a major role in marsh vegetation, the muskrat (*Ondrata zibethica*). Muskrats are rodents that have an impressive capacity to reproduce. They have six or more young per litter and under optimal conditions, can have four litters per year. Muskrat populations have been reported to grow so rapidly that they literally "eat out" the plants from the marsh (Weller 1981).

The muskrats use the plants for food and construction. The rootstocks, rhizomes, and tips of the plants are eaten and the stems are used to contruct a plateform and left to rot. In the early stages of reestablishment in a marsh, muskrats tend to construct lodges nearly 2 m high and 4.5 m in diameter. However, the lodges are half that size when the muskrat population has removed a large proportion of the plant material. Reproduction rates have also been reported to undergo changes that reflect the conditions of the marsh. Thus, as the marsh becomes crowded, lodges decrease in size, and the number of surviving young decreases. These dynamic activeties affect the numbers and species of birds which inhabit a marsh. The greatest number of individual birds are supported when the whole marsh is dominated by a thick growth of emergent plants. However when there is a roughly equal balance between open water and emergent vegetation, then the maximum number of species can be supported by the marsh. When muskrats open up the vegetation around their lodges, they create habitat diversity that attracts an increased diversity of bird species (Weller 1981).

The most extensive swamps in North America occur along the Atlantic and Gulf coastal plains. The most notable examples include the Great Dismal Swamp, Okefenokee Swamp, and Everglades. The Great Dismal Swamp consists of nearly 200 km^2 witnin the Atlantic coastal plain, The swamp is believed to be at least 80.000 years old (Kirk 1979). The oldest pollen is from pines and spruce. The modern vegetation consists of a variety of tree species including cypress, white cedar, and tupelo gum (*Nyssa aquatica*). The swamp acts to stabilize the groundwater table in the region. During periods of very heavy rains, the swamp controls floodwaters by temporarily storing excess runoff and releasing it slowly as runoff or into the groundwater system. The area abounds with an amazing variety of plants and animals, some of which are pooly known (Kirk 1979).

Within the Everglades of south Florida and the Gulf Coast swamps, one organism has had an especially important role within the environmental system, the American alligtor (*Alligator mississippiensis*). Alligators are active diggers. They dig out channels and build large mounds for nest sites. They also dig deep pools or "gator holes" in the swamp. During periods of extreme drought these pools provide the last remaining inhabitable sites in the swamp for birds and other animals to survive. When the drought breaks repopulation of the swamp takes place from the animals which have survived at gator holes. The nesting mounds of alligators contribute to the variety of the swamp environment by providing nest sites for several bird species, such as herons and egrets. Thus, marshes and swamps are complicated habitats that depend on energy expenditure and behavioral adjustment of specialized water-dwelling animal inhabitants. The activities of these animals create conditions that provide opportunities for existence of the larger plant and animal community.

Beaver Impoundments

Beavers are a *sine qua non* of the influence of animals on streams. They have an unusually large impact on stream systems and the development of valley soils and vegetatin of the Northern Hemisphere. Two species are recognized, the Eurasian beaver, *Castiber*, and the North American beaver. *C. canadenss* The beaver is a large rodent, second in size to the capybara (*Hydrochoerus hydrochaeris*), a relative of the guinea-pig and inhabitant of tropical America. The beaver attains a length of between 90 and 120 cm and a weight of up to 30 kg. The present range of the beaver is only a fraction of its former range. In the Old World the range of *C. fiber* once extended from Scandinavia to Spain and from Britain to Siberia. Now there are only limited populations in the lower Rhone, the middle Elbe, and the southern part of Norway. The former range of *C. canadensis* included the northern half of North America with extensions into the southern Appalachians and the southern Rocky Mountains. They numbered in the thousands of millions and were an attraction to early traders who sought their pelts. In general, beavers were hunted for food (because of the scales on their tails they were defined as fish and could be consumed by Catholics on Friday and during Lent), for fur (they have magnificent coats which are durable and easily tailored). and for castoreum, a scent gland substance which was highly valued for its medicinal properties during medieval times. For these reasons and because they occupied flood plain soils that humans valued for agriculture, the populations of beavers were cut to a fraction of their former number by hunters and trappers. But, many soil and vegetation features can be understood by only considering the influence that beavers have on river valleys.

The activities of the beaver are entirely bound up with trees, flowing water, and the construction, of dams Beavers create deep ponds or maintain water levels in preexisting depressions by building dams across streams. These are usually made of logs and sticks, although dams made of boulders have also been described (Chapman & Chapman 1937). The impoundment stores water and serves the beaver by providing a site for a lodge which is safe from predators. The food of the beaver consists of the bark and shoots of popar, willow, alder, elm, or oak, plus vegetation which is favored by the expanded riparian environment that the beaver pond creates. The beaver obtains food by chopping down trees. The bark is stripped and the construction of the beaver dam or the lodge which is built in the impounded waters behind the dam. In order to move large logs, beavers construct canals.

Castor canadensis is acknowledged as the superior dam builder. The largest recorded dam built by *C. fiber* was found in the Voronezh region of Russia. It was 120 m long, 1 m high, and 60—100 cm wide. One of the largest structures constructed by C. *canadensis* was on the Jefferson River in Montana: it was 700 m long and could bear the weight of horses and their riders. The largest dams were probably maintained by generations of beavers for centuries. Beavers regulate the water level of the beavers pond by constantly adjusting the dam. During high water, the dam can be adjusted by lowering the crown. During low water the beavers tighten up the dam to reduce seepage. The splendid team work beavers building and maintaining a single dam is itself magnificent. A series of dams within a number of tributaries of a drainage basin operate together to create an impact on the landscape.

The concerted efforts of many beavers must have changed many valleys of the Northern Hemisphere during the Holocene from relatively physically and biologically homogeneous habitats into heterogeneous environments supporting diverse groups of organisms. The process can be imagined within the high valleys of the Rocky Mountains. Prior to the movement of beavers into high valleys, conifers probably dominated the landscape. As beavers move in and construct dams, the impounded waters flood out the conifers. Each pond creates a site which is attractive to migrating waterfowl and other birds. The riparian environment of the shores of the pond supports new types of plants, that supply the food for the beaver population. The pond checks the movement of sediments down stream and these particles add to the richness for aquatic organisms of the pond. The enlarged surface area of the enriched pool of water supports high levels of photosynthesis and releases increased amounts of organic carbon that in turn enhances the food supply for aquatic organisms down stream. Eventually, the impoundment becomes filled with sediments, and baevers abandon the site. As the beaver pond drains it becomes a thick lens of rich humus that supports the rapid growth of grasses and forbs (non-grass species) and establishes the beaver meadow. These soils have high water-holding capacity and slowly release water to the water table and the stream. The activity of the beaver becomes the primary force in the evolution of the soils in the riparian zone. They influence the diversity of habitats and establish structures which augment the slow but steady movement of water through the drainage basin. These feats give the beaver its reputation as the unparalled engineer and builder of animal kidgdom (Hodgdon & Larson 1980).

The estuary

Eventually, most freshwater streams flow to ocean level and meet the sea. Under special conditions, the flow of fresh water to the ocean is parttally enclosed and the outlet to the sea restricted to a relatively narrow opening. These places are referred to as estuaries. The most important estuaries include drowned river drowned valleys, such as Chesapeake Bay ; down faulted sections of the Earth's crust, such as San Francisco Bay ; drowned glacial troughs, or fjords, found in Norway, Alaska, Nhe Zcaland, and other glaciated mountainous coastal regions ; and saltmarshes and lagoons that are formed behind barrier beaches.

Within the estuary, fresh water from the land flows upon and meets the denser salt water of the ocean. The interface between the two different types of water is a nutrient-rich and highly productive site for organisms. But these sites are also physically demanding and only a few species can tolerate the rigors of successive exposure to fresh and sea water, and the wide fluctuation in temperature associated with these environments.

One component of the estuary is shallow saltmarshes that play an important role for organisms. Saltmarshes are regularly rinsed by the ebb and flow of the tides. Waste products are diluted and removed and nutrients are brought in from the land and the sea. The materiais of streams flocculate as the mixture of clay-size particles and algae meets the sea waters. They settle out, leaving the water clear (Avnimelech *et. al.* 1982). In the saltmarsh, plants need not expend energy in collecting minerals. They are assisted in obtaining minerals from many organisms. Clams, oysters, and muscels trap nutrients as they feed, and deposit the wastes as pseudo-feces, In these nutrient-rich waters, plants grow, die, and cycle rapidly as they are consumed by the inhabitats of the estuary. The base of the food chain of the estuary is marsh grass or *Spartina*. The major energy flow is by way of the detritus food chain rather than the grazing food chain. In estuaries along the coast of Georgia, U.S.A., for example, *Spartina alterniflora* dominates, and organic detritus (over 90% of which orginates with the growth of *Spartina*) is the link between primary production and consumption. Only a small portion of the grass is consumed as a part of the grazing chain (Ranwell 1972).

Estuaries are one of the most productive habitats on Earth. They provide huge food supplies as well as sanctuary for many species of waterfowl. They serve as a breeding ground and nursery for many marine fish species. They store nutrients for larval and juvenile marine life. As various species mature, they leave the estuary and head out to sea. Thus, there is a constant movement of the rich organic and nutrient materials to the sea (Honjo 1982). As tides move in and out there is also a movement of the materials out to sea. And thus it is that the stream flows to the sea and enriches the shallow coastal waters of the marine environment.

Conclusions

In this chapter, the hydrologic cycle, and especially the terrestrial ramifications of the cycle as seen in the organization of the drainage basins and the mayriad of organisms that exist along the path of the stream were examined. The path of water from the atmoshpere to the land, and across or through the soils to the stream and then through a number of specialized stream, marsh and pond communities within the flow of water were described. At all points, the climate and geologic setting provide primary conditions, but organisms moderate and alter the flow of water, minerals, and organic carbon of the stream. Thus, we have followed the flow of water from the atmoshere to the soils, from the soils to the stream, and along the streams to lakes and wetland communities and finally to the estuary. At every juncture of the path from the land to the sea the stream supports thriving group of organisms.

5

Sedimentary Environments

In the preceding chapter, our review of nonmarine dipositional settings took us on a downhill journey starting with the alluvial fan that forms at the foot of a mountain and ending in a lowland river valley. In effect, we moved down the slope of the hypsometric curve, the graph which depicts the percentages of the earth's surface lying at various position above and below sea level. We will now continue this downhill trip, first examining sedimentary environments along the edge of the sea and then moving toward the deep ocean floor.

Deltas

If a river does not flow into a larger one, it usually either into a lake or into the sea. In doing so, its current dissipates, and it often drops its load of sediment in a fanlike pattern. The depositional body of sand-silt, and clay that is formed in this way is called a **delta** because of its resemblance to the Greek character. Most of the large deltas that have been well preserved in the geologic record formed in areas where sizable rivers emptied into acient seas.

Delta-plain beds, which consist largely of sand and silt, are nearly horizontal except where they are locally crossbed. Some delta-plain deposits accumulate within channels. As a river slows down on the surface of the delta, sand builds up on the bottom, causing the river channel to branch repeately into smaller channels that radiate out from the mainland. These **distributary channels** are floored by cross-bedded sands. Sand also spills out from the mouths of the channels, forming shoals and sheetlike sand bodies along the delta front. Between the distributaries and separated from them by levees are swamps, which are sometimes dotted by lakes. Here, as in the overbank areas of meandering rivers, muds accumulate and marsh plants of ten grow, contributing to future coal deposits.

Delta-front (or foreset) beds slope seaward from the delta plain, usually lying in waters that are deeper than those agitated by wind-driven surface waves. These bends consist largely of silt and clay, which can settle under these quiet conditions. Since they lie fully within the marine system, delta-front muds harbor marine faunas that often leave fossil records, but these muds usually contain fragments of waterlogged wood as well. In fact, the presence of abundant fossil wood in ancient marine muds testifies to the presence of both land and a river system near the site of preservation. In short, most ancient subtidal muds that harbor abundant fossil wood debris represent deltaic deposits.

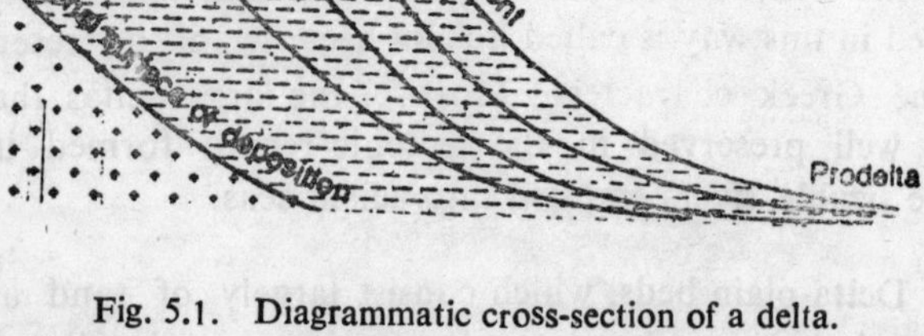

Fig. 5.1. Diagrammatic cross-section of a delta.

Spreading seaward at a low angle from the lowermost delta-front deposits are the prodelta (or bottomset) beds, which consist of clay. Even during floods, the freshwater that spreads from distributary channels slows downs so rapidly that it is partly because freshwater is less dense than saline marine water that clay is carried far from distributary mouths. Because the freshwater floats on top of the more dense marine water, it does not mix in quickly; instead, it spreads seaward for some distance, carrying much of its clay.

As a delta **progrades**, or grows seaward, relatively coarse delta-plain deposits build out over finer-grained foreset beds in accordance with Walther's law. These delta-front beds build out over still finer-grained bottomset beds, resulting in a sequence of deposits that coarsens toward the top. As we will now see, the stratigraphic expression of this upward-coarsening sequence varies in accordance with the nature of the delta. We will first consider the nature of a delta that grows into a relatively quiet sea and we will then examine one that grows seaward in the presence of strong waves.

The famous Missisippi River delta is currently building out into the Gulf of Mexico in an area that is protected from strong wave action. As a result, this delta projects far out into the sea. Because construction of the delta from riverborn sediment prevails decisively over the destructive forces of the sea, this type of delta is sometimes called a **riverdominated** delta.

That protion of the Mississippi which is growing at any given time is much smaller than the delta as a whole. This growing portion, or **active lobe**, is the site of the currenly functioning distributary channels. Many previously active lobes can identified in the delta-plain portion of the Mississippi delta, and these lobes, which have been dated by the carbon 14 method provide a history of deltaic development during the past few thousand years. It is known, for example, than the site of depositional activity (or lobe growth) periodically shifted in this delta when floods caused the river to cut a new channel and to abandon the previously active channel and its distributaries.

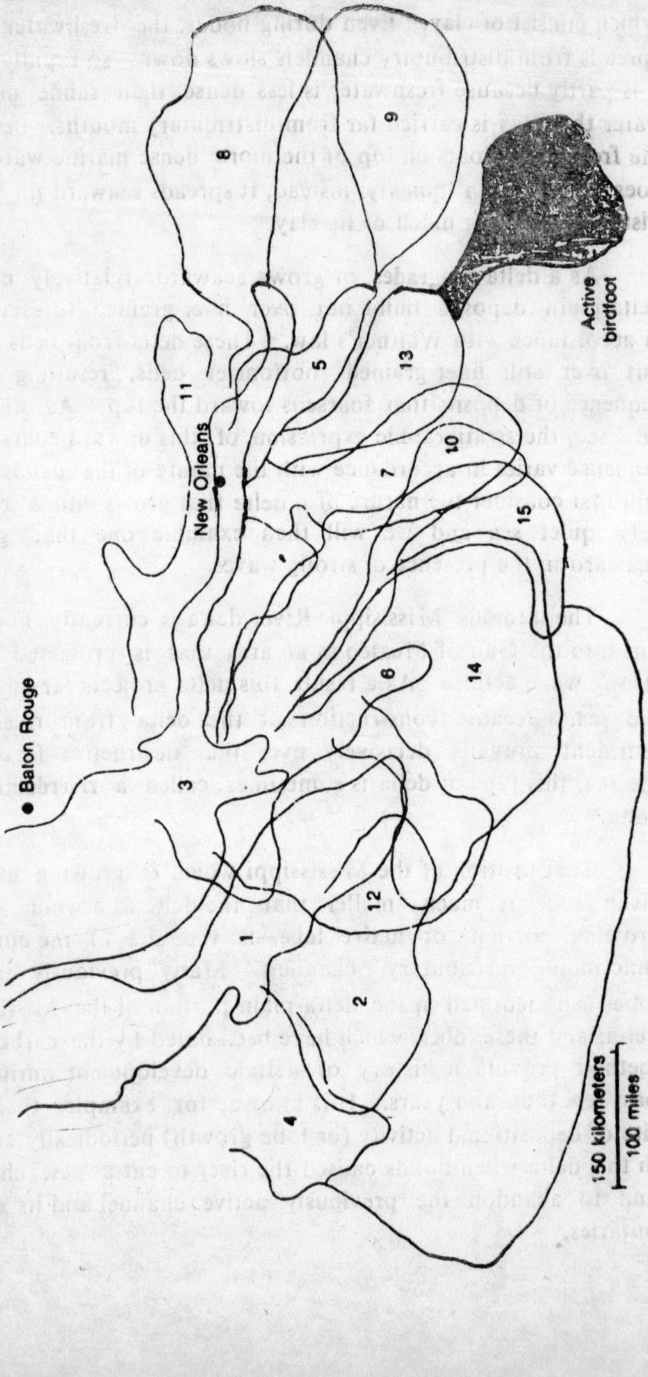

Fig. 5.2. Lobes of the Mississippi delta.

The fate of an abandoned delta lobe is, in fact, the key to understanding the stratigraphic sequence produced by a river-dominated delta. In short, an abandoned lobe gradually sinks. It does so for two reasons; first, the sediments from which it is formed compact under their own weight; and second, the lobe is part of the entire delta structure which is also constantly sinking as a result of the isostatic response of the underlying crust to the weight of the continually growing mass of sediment. After an abandoned lobe settles, a younger lobe will eventually grow on top of it. Each lobe will then consist of the typical upward-coarsening sequence. The result is an accumulation of cycles that differ markedly from those of meandering rivers, which, it will be recalled,

Some deltaic cycles in the rock record lack tops or bottoms as result of erosion that occurs before another cycle is superimposed. Because the delta-building activity of a river-dominated bed is limited to the active lobe at any given time, beds representing such a delta can seldom be traced very far laterally in the rack record.

Two African rivers—the Niger, which flows into the Atlantic; and the Nile, which flows into the Mediterranean—are currently forming deltas that have been less successful than the Mississippi in overcoming the destructive forces of the marine. realm. These **wave-dominated** deltas are currently forming in areas where wave activity is so strong that it prevents a small active lobe from growing out into the tea. Instead, much of the delta as a whole is traversed by active distributory channels, and the sand that these channels transport to the sea is quickly spread to form beaches and transverse bass. Seaward of these ise a zone of silt that grades into finer-grained mud. In effect, ths entire delta consists of a single active lobe. This lobe prograde as a unit, representing one large, upward coarsening stratigraphic sequence.

Ancient deltaic deposits that consist of many cycles are important sources of petroleum, which tends to become trapped in the topset sands of one cycle beneath the impermeable muds at the base of the succeeding cycle. Many buried Cenozoic deltas bordering the Gulf Coast, for example have yielded large quantities of petroleum.

The Barrier Island-Lagoon Complex

Because deltas form exclusively at River mouths, they occupy only small percentage of the total shoreline of the world's oceans. Much longer stretches of shoreline are fringed by **barrier islands**, which are linear structures composed largely of clean sand that has been pied up by waves. Although there are barrier islands that extend laterally from some wave-dominated deltas, most barrier islands derive their sand not from neighboring rivers but from the marine realm itself. They are built by shallow currents called **longshore currents**, which winnow sediments and sweep sand parallel to the shorline.

Shallow **lagoons** lie behind long barrier islands such as those that border the Texas coast of the Gulf of Mexico. Because they are protected from strong waves, these lagoons act as traps for fine-grained sediment and are usually floored by muds and muddy sands. River that empty into lagoonsof ten build small deltas.

Freshwater from rivers and streams tends to remain trapped in coastal lagoons for some time. Thus, lagoonal salinities in moist climate are often brackish, changing during the year with the changing rate of freshwater runoff from the the land. Laguna Madre of Texas is typical of the lagoons found in more arid climates, where temperatures are high. Because they receive little freshwater from rivers and suffer a high rate of evaporation, the ponded waters of this long lagoon are hypersaline. Whether they are brackish or hypersaline, however, the abnormal and fluctuating salinities of lagoons exclude many forms of life that require normal marine salinities. As a result, ancient lagoonal sediments are characterized by fossil faunas of low diversty. Those species that are present in lagoons, however, often occur in large numbers, and some of them are usually burrowers that disturb the muddy sediments the lagoonal floor, leaving those sediments either mottled or homogeneous and largely devoid of bedding structure.

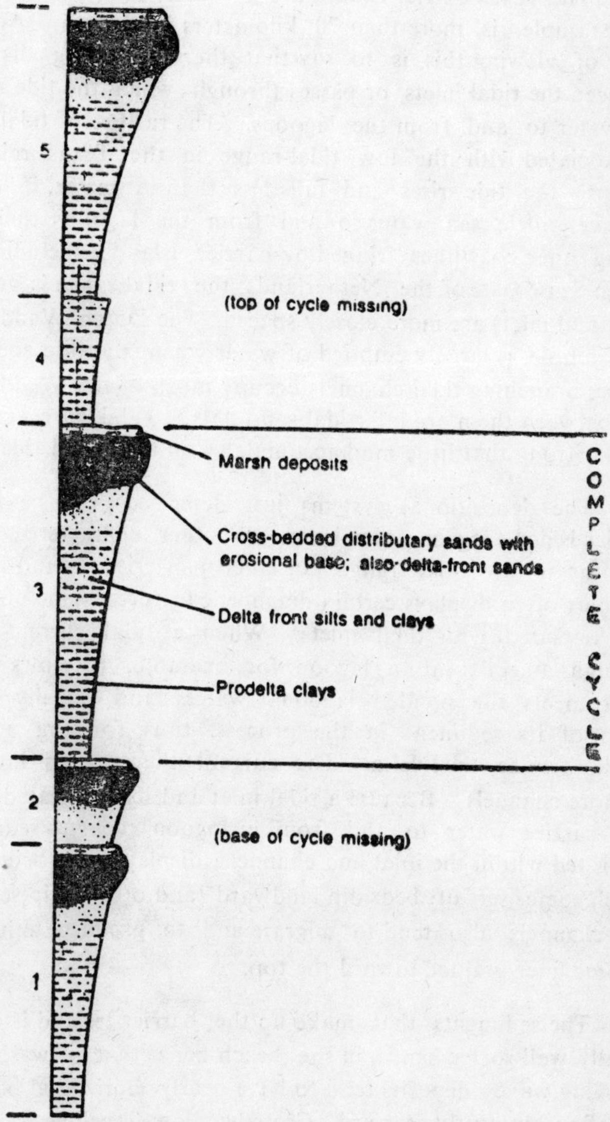

Fig. 5.3 An idealized stratigraphic section representing five detaic cycles.

The Texas barrier islands are unusually long. Padre island, for example, is more than 70 kilometers in length. Another way of viewing this is to say that there are long distances between the tidal inlets, or passes through which the tide moves sea water to and from the lagoons. The rarity of tidal inlets is associated with the low tidal range in the Texas reigon—because the tide rises and falls less than a meter, it moves relatively little sea water to and from the lagoons each day. Along other coastlines fringed by barrier islands, including the North Sea coast of the Netherlands, the tidal range is greater, and tidal inlets are more closely spaced. The Dutch Waddenzee, for example, is largely emptied of water when the tide goes out. Large, branching tidal channels occupy much of the Waddenzee, and between them are intertidal sand flats. Tidal currents here are so strong that little mud accumulates on the broad tidal flats.

The depositional systems just described can be called **barrier island—lagoon complexes.** Whether characterized by a high or a low tidal range, an ancient barrier island-lagoon complex often displays certain diagnostic features, some of which are associated with tidal inlets. When a tidal current flows through an inlet into a lagoon, for example, it slows down as it meets the ponded lagoonal waters and usually drops some of its sediment in the process. thus forming a structure known as **tidal delta.** The current then divides into two or more channels. Because a tidal inlet and its associated channels carries water to and from a logoon, coarse sediments deposited within the inlet and channels display cross-bedding in which some sets of beds dip landward and others dip-seaward The channels also tend to migrate and to produce units that become finer-grained toward the top.

The sediments that make up the barrier island itself are usually well-sorted sands, in the beach zone that is washed by breaking waves, deposits tend to have nearly horizontal bedding but often dip gently seaward. Cross-bedding develops in areas where the beach surface is gently irregular and changes from time to time. Wind-blown sand often accumulates behind the beach as sand dunes, but these dunes are often eroded before burial and are thus missing from many depositional sequences that represent barrier island lagoon complexes.

A number of sediments are found along the shores of lagoons. Among these are **tidal flats** formed of sand or muddy sand whose surfaces are alternately exposed and flooded as the tide ebbs and flows. High in the intertidal zone, above barren tidal flats, marshes often fringe one or both margins of a lagoon. Here the high rate of production of plant debris leads to the accumulation of peat, which can become coal after a long period of burial.

When a barrier island-lagoon complex receives sediment at a sufficiently high rate, it progr ades—that is, it migrates seaward—like the active lobe of a delta. Unlike the migration of a delta, however, this progradation takes place along a broad belt of shoreline. Here, as the shoreline migrated eastward, marsh and tidal flat deposits prograded over sediments of the lagoon and associated tidal channels. These, in turrn, built out over deposits of the barrier beaches and over the tidal deltas and marshes behind them. Thus, the horizontal sequence of environments came to be represented by a corresponding vertical sequence of sedimentary deposits in accordance with Walther's law.

If erosion occurs during environmental migration, some of the deposits of a particular environment may be missing from a vertical sequence. However, despite the frequent imperfection of the record it leaves behind, the barrier island-lagoon system illustrates particularly well how the close vertical association of sediments that represent adjacent depositional environments can be used to identify ancient depositional systems. We have seen vertical associations of rock types are also very useful in the identification of ancient meandering-river deposits and deltas.

Off shore shelves and inland seas

Of course, many depositional environments intergrade rather than remaining sharply separated from one another in he manner of marshes and lagoons. As a result, the sedimentary deposits thus produeed often integrade vertically in the geologic record in accordance with Walther's law. We have already seen, for example, that the prodelta clays of a deltaic sequence usually grade upward into silty sediments of the delta front. There is often a similar upward-coarsening gradation in the record left by a migrating barrier island because barrier-island because usually grade into finer sediments in a seaward direction. Because these finer sediments represent progressively deeper environments of a continental shelf or a shlalow inland sea, they are deposited on continental crust rather than in the deep sea. When a barrier island progrades, it builds out over the finer-grained deposits that represent the deeper water in front, resulting in an upward coarsening sequence below the package of sediments that represents the beach and lagoon.

More details of this type of sequence are which illustrates the vertical profile that would result if progradation were to occur in the modern Gulf of Gaeta, in Italy. Although the upward-coarsening sequence is evident, other typical biological and physical structures in the sediment also display a characteristic vertical order. Very few a animals for example, are able to live in the shifting sands at the edle of wave-ridden beaches, where physically produced bedding structures are much more common than biologically produced burrows. But farther offshore, below the lower limit of wave activity, burrowers are more common, and physical processes have little effect except during major storms; consequently, burrow structures are more abundant in this sediment than are structures produced by physical processes. Thus in a profile representing progradation of beach sands over offshore muds, biologically produced sedimentary structures predomminate near the bottom, while physically produced structures prevail toward the top.

Strong tidal currents that sweep over continental sheives can form sand bars well below the limit of wave motion. It is presumedhowever, that more confined inland seas, which are largely cut off from the immense tidal movements of the open ocean, usually experience weak tidal currents. Unfortunely, we have few island seas to examine in order to verify this assumption, since the level of the ocean is lower today than it has been at most times in the past. The study of sedimentary rocks formed in ancient inland seas does indicate, however, that the sandy deposits that usually fringed the seas generally gave way to muddy sediments in deeper water.

Organic Reefs

In tropical shallow marine settings, if siliciclastic sediments are in short supply, carbonate sedimentation usually prevails. Here, coral reefs are often prominent features. Because they are produced largely by organisms that secrete calcium carbonate, organic reefs form their own distinct depositional records—as bodies of limestone.

A modern reef has several structural components that form a heterogeneous limestone deposit. The basic framework of a reef consists of the calcareous skeletons of organisms, primarily corals. This framework is strengthened by cementing organisms, including coraline algae, that encrust the surface of the reef. Carbonate sediment, which is composed of fragments of the skeletons of reef-dwelling organisms, is trapped within the porous framework, filling many of the voids. With their complex internal structure, reef limestones are for the most part either unbedded or only poorly bedded. Even with the presence of infilling debris, reef limestone is highly porous. This accounts for the fact that many ancient buried reefs serve as traps for petroleum, which migrates into them from sediments rich in organic matter.

Because living reefs stand above the neighbouring sea floor, they alter patterns of sedimentation nearby. On the leeward side of an elongate reef, there is often a lagoon where waters tend to be relatively calm. This is especially true if the reef has a typical **reef flat**, or horizontal upper surface, that stands very close to sea level. Below the living surface of the reef is a limestone core consisting of dead skeletal framework, dead skeletons of camenting agents, and trapped sediment. Extending seaward from the living surface, there is often a pile of rubble called **talus**, which has fallen from the steep, wave-ridden reef front.

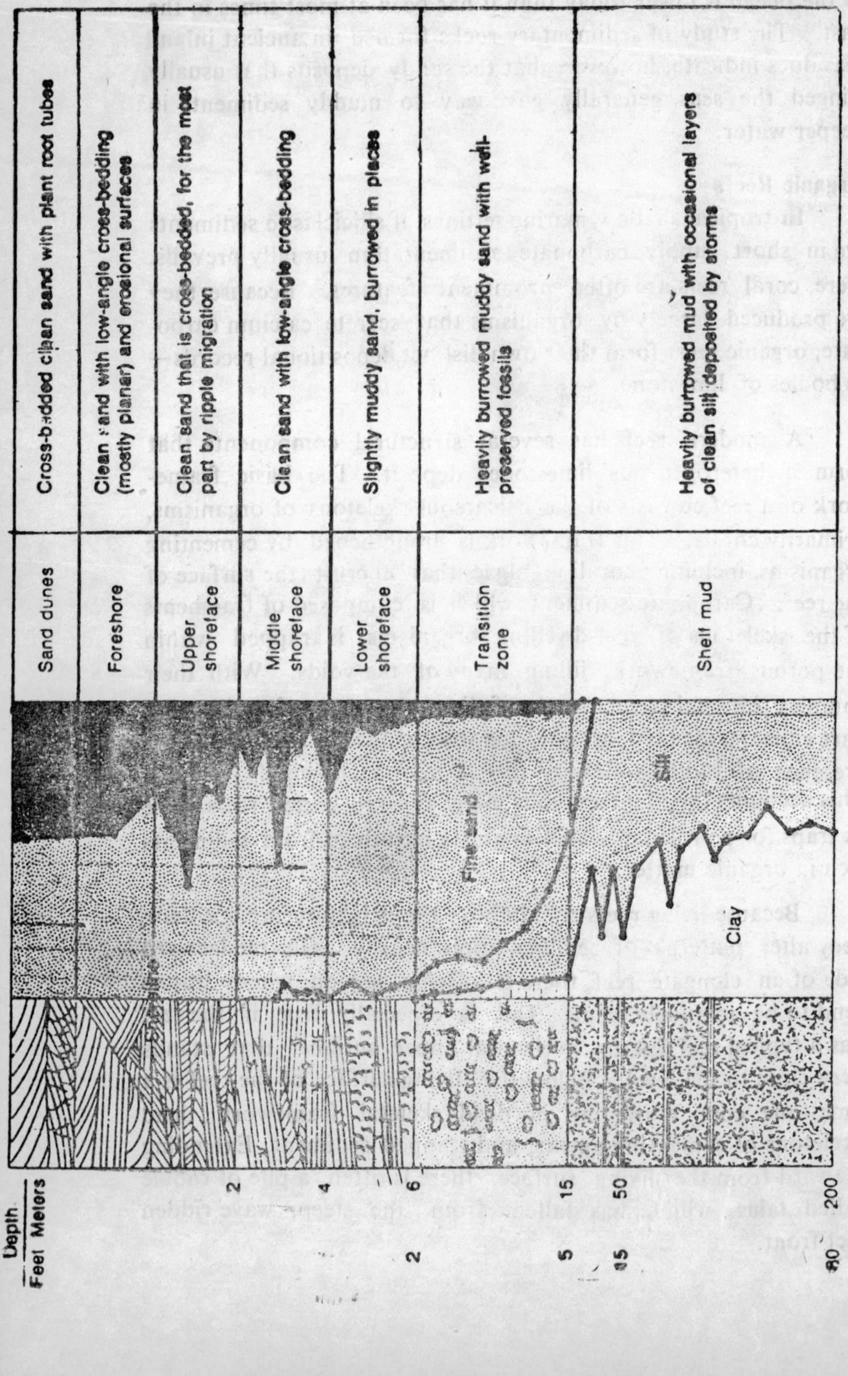

Fig 5.4. A hypothetical stratigraphic sequence that would result from progradation at the Gulf of Gaeta.

Most reefs have the capacity to grow rapidly and to keep pace with rising sea level or a subsiding sea floor. Reefs build upward rapidly enough to remain near sea level even when the sea floor around them is becoming deeper. Many reefs, in fact, grow so rapidly are and so durable that they build seaward in the manner of a prograding delta. Although it was built by organisms that have been extinct for hundreds of millions of years, this reef closely resembles many modern reefs in its basic structure—it displays both a seaward talus slope and a leeward reef flat.

As water flows over a reef flat, it is slowed by friction. This gradual dissipation of the energy of motion results in what is known as an **energy gradient**. Because different species of framwork builders and cementing organisms thrive under different energy conditions, a zonation of species occurs across a typical reef flat from the coraline algae that form a seaward ridge to species that flourish best under the very quiet conditions in the rear. The living reef adjacent to Bikini Island in the South Pacific illustrates these biological zones, which run parallel to the reef front Isolated **patch** or **reef spinnacle reefs** are often found in lagoons behind elongate reefs. Surrounded by quiet water, these small structures tend not to display zonation. Elongate reefs that face the open sea and have lagoons behind them are known as **barrier reefs**. Other reefs grow right along the coastline without a lagoon behind, and these are known **fringing reefs**. Some fringing grow seaward and eventually become barrier reefs.

Perhaps the most curious reefs in the modern world, however, are the circular or horseshoe-shaped structures known as **atolls**. Atolls frequently form on volcanic Islands and are thus quite common in the tropical Pacific, where such Islands are common. Charles Darwin offered an explanation for the origin of the Pacific atolls that is still accepted today. According to Darwin, each atoll was formed when a cone-shaped volcanic Island was colonized by a fringing reef. The island eventually sank beneath the sea, leaving a circular reef standing alone with a lagoon in the center, where limestone now accumulates in quiet water. Often the reef does not quite form a full circlo but is instead broken by a channel one the leeward side, where reef-building organisms do not thrive. Horseshoe-shaped atolls range up to about 65 kilometers (40 miles) in diameter during World War II, they served as natural harbors for ships.

Ancient atolls that lie buried beneath younger sediments can be identified from the study of cores of sediment brought up from drilling operations—and because porous reef rocks often serve as traps for petroleum, drilling if the vicinity of these atolls is often a profitable venture.

Carbonate Platforms

A carbonate platform is a structure that is formed by the accretion of carbonate sediment and that stands above the neighbouring sea floor on at least one of its sides. Like organic reefs, carbonate platforms consist largely of calcium carbonate that originated in shallow tropical waters at or near the site of accumulation. In fact, organic reefs form parts of many carbonate platforms, often growing along the windward margins of these platforms.

In times past, carbonate platforms have stretched along most or all of the earstern margin of the United State—but because climates are cooler today than they have been during most of the earth's history, these structures are not well represented in the modern world. The tropical areas of the modern world do offer several examples, however in the western Atlantic, for example, a large carbonate platform extends seaward from the Yucatan Peninsula of Mexico. There are also smaller platforms bordering the islands of the Antilles, and the platforms known as the Little Bahama Bank and Great Bahama Bank lie to the east and southeast of Florida. Because the varied sediments that are now accumulating on the Bahama banks resemble those of many ancient carbonate platforms, a review of these sediments will serve to introduce carbonate sediments in general.

One general property of carbonate platforms that the Bahama Banks illustrate is rapid deposition of sediments. Since mid-Jurassic time, about 170 million years ago, some 10 kilometers (−6 miles) of shallow-water carbonates have accumulated both on the Bahama banks and in southern Florida, which was part of the same carbonates platform during Cretaceous and earlier times. The heavy buildup of carbonate sediments in this region has caused the oceanic crust to bend downward, and shallow-water Jurassic deposits now lie up to 10 kilometers below sea level. During the Cenozoic Era, the termination of carbonate deposition in some areas led to the development of the deep channels known as the Straits of Florida, Tongue of the Ocean, and Exuma Sound. These channels now separate the Florida peninsula, the Little Bahama Bank, and the Great Bahama Bank.

Sediments in the vicinity of Andros Island, the largest of the Bahamas, include a variety of important carbonate types. Trade winds from the east-northeast provide water currents and nutrients that are highly favourable to the growth of coral reefs east of Andros Island, which borders Tongue of the Ocean. In the wind shadow of Andros, muddy carbonates accumulate. in a quiet, shallow lagoon. These consist primarily of aragonite neddles derived from the breakdown of the skeletons of calcareous algae that inhabit the floor of the lagoon. Large quantities of these muddy sediments occur as **fecal pellets**, oval grains composed of mud that has been bound up with mucus after passing through the guts of deposit-feeding worms or other animals. Some of these bellets have become cemented on the sea floor and have thus developed into grains. Many lagoonal deposits that have been studied in the rock record also consist of pelleted and unpellted lime muds such as those that floor the quiet, shallow seas west of Andros Island.

Outside the wind shadow of Andros Island, sediments of coarser grain size predominate. The most spectacular of these sediments are clean sands called **oolites**, which consist of concentrically layered, nearly spherical grains. Each individual grain, or **ooid**, forms by accretion—as the grain rolls around on the sea floor, aragonite needles accumulate tangentially on its surface. This rolling action is essential for spherical grain growth, which is why ooids tend to form in areas where there is strong wave activity. In the Bahamas, ooids occur primarily along margins of the Great Bahama Bank, where tidal currents are strong. In the clear, shallow water of these margins, ooids form magnificant submarine dunes. Oniy 100,000 years ago, when the seas stood slightly higher than they do today, ooid dunes extended along the southern Florida coastline. Cross-bedded oolites that occur in the rock record represent shallow environments affected by strong tides.

Andros Island is friged on its leeward (western) side by intertidal carbonate deposits that have numerous counterparts in the rock record, These intertidal sediments consist primarily of pelleted muds that have been washed up by storms from the subtidal sea floor on which they formed. The intertidal pellets are hardened to varying degrees by calcium carbonate cement, and they are accompanied by a much smaller volume of skeletal grains.

Still other features of Andros Island characterize carbonate platform environments in general. The intertidal zone west of the Island, for example, is bordered by a supratidal beach ridge composed of material that has been thrown up by storm waves. Here, tidal channels are bordered by supratidal levees similar to those found along meandering rivers on the land. Beyond the supratidal levees are marshy tidal flats covered by blue-green algae and by sparse mangrove bushes. Shallow ponds also occupy some of the areas between tidal channels. Island from this intertidal complex is a supratidal march, which also consists of blue-green algae and mangrove shrubs.

The seaward-facing slopes of the supratidal breach ridge are wetted periodically and then dried and cemented, causing the surface sediment to form a weak crust. In many places, the crest of the beach ridge is surfaced by partially overlapping diagonally stacked, flat pebbles which were washed in by storms that ripped them from the seaward-facing slopes. Such diagonally stacked pebbles characterize supratidal settings in areas of carbonate deposition and are frequently found in ancient carbonate rocks.

BorderIng the Andros tidal channels in some places are knobby intertidal structures known as **stromatolites**. which are produced by threadlike algae that form sticky mats, Repitition of this process on an irregular surface forms a cluster of stromatolite. Stromatolites are produced by blue-green algae, which, are among the most primitive groups of organisms in the world. The fossil record of stromatolites is unusually ancient, extending back through more than 3 billion years of geologic time.

On the supratidal backslope of the Andros beach ridge are other layered sediments that are also formed by a succession of algal mats. In the supratidal beach ridge setting each algal mat is formed at a time of flooding by storm waves. In this setting, the surface of accumulation is relatively flat, and the layers are not domed into stromatolites. The horizontal layering of the sediments, together with numerous holes transecting the various layers, gives these sediments the informal name bird's eye limestone. Although some of the holes in this sediment are tubes produced by burrowing worms, others represent spaces that have been left where strands of algae have decayed, gas bubbles have formed, or sediment has dried and cracked Currently, well-ayered bird's eye limestones occur primarily in supratidal and high intertidal settings, when found in the rock record, they usually represent such settings.

There is a simple reason that stromatolites and bird's eye limestones are found almost exclusively in supratidal and high intertidal settings. Because these environments are exposed above sea level much of the time, they become hot and dry, and relatively few marine animals can survive in them. Thus, there is little to interfere with the tendency of algal mats to form layered structures. In subtidal settings, algal mats grow here and there, but they are eaten by the many grazing animals and

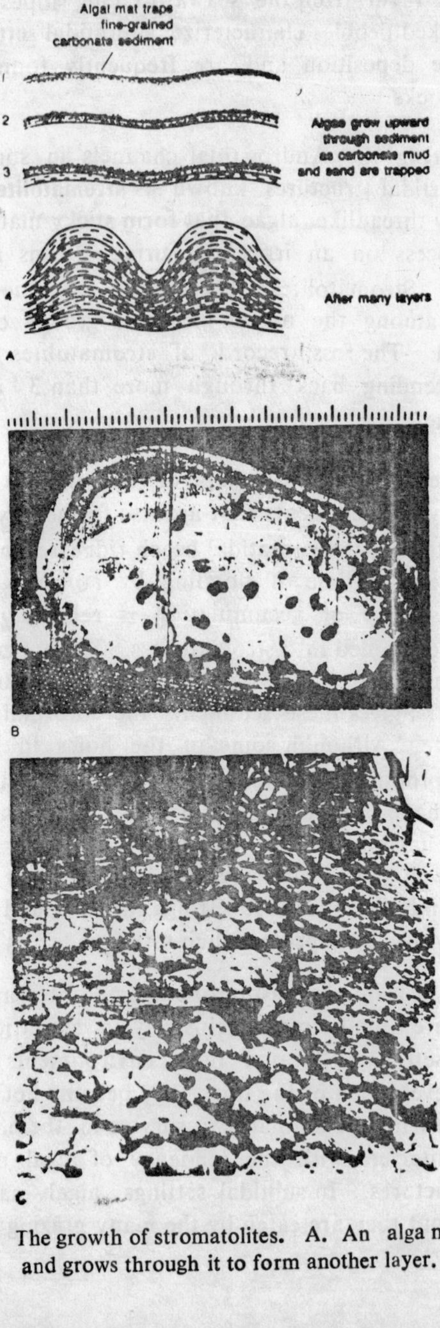

Fig. 5.5. The growth of stromatolites. A. An alga mat traps mud and grows through it to form another layer.

damaged by burrowers : as a result, they do not survive well below water and seldom accumulate to form stromatolites or well-layered limestones. As we will some very ancient stromatolites may have formed subtidally before the origin of animals that feed on filamentous algae.

The supratidal backslopes of tidal channel levees adjacent to Andros Island dry out after occasional flooding by some-driven seas. As a result, the surface of the sediment here, which is also bound by an algal mat is broken by mud cracks resembling those that often form when a mud puddle on the land dries up. Mud cracks associated with ancient marine deposits usually represent intertidal or supratidal environments that were alternately wetted by the sea and dried by the sun.

In general, then, a close association in the rock record to flat-pebble conglomerates, stromatolites, well-layered bird's eye limestones, and mud cracks points to a complex of intertidal and supratidal carbonate platform environments. If such deposits are prominently displayed in a carbonate sequence that is hundreds or thousands of meters thick and that represents millions of years of deposition, it is almost certain that they formed on a slowly subsiding carbonate platform resembling the Bahama banks.

Marginal Marine Sabkas

Shallow-water carbonate sediments do not accumulate only on platforms that stand high above the surrounding sea floor. Sometimes they fringe the land along gently sloping shelves, as is the case along the southern margin of the Persian Gulf. Here, barriers separate elongate lagoons from open marine water. These barriers consist of three kinds of structures : islands with dunes, submarine ooid shoals, and coral reefs. Muddy carbonate sediment accumulates in the quiet lagoons, and inland from these are broad intertidal and supratidal flats.

The intertidal sediments of the southern Persian Gulf are well-lay('ed algal-mat deposits that grade into a sabka. A sabka is a flat supratidal area heavily laden with evaporites in the form of nodules, crusts. and crystalline aggregates, halite, anhydrite, gypsum, and mineral grains resembling dolomite are all present. Some of these evaporites from because the arid climate causes marine waters that soak through from the tidal flat and lagoon to evaporate rapidly, however, at least some of the evaporating waters seep into the flat from highlands to the south.

Because of rapid deposition, the supratidal and intertidal sediments of the southern Persian Gulf encroach on the lagoon. As a result, cores obtained by drilling in the supratidal sabka pass down through the sand and evaporite layer that is forming here to algal laminations of the tidal flat, which previously extended inland to this position. Below these are lagoonal sediments, whose presence reflects a time when the shoreline was still further inland. In other words, as the sediments push the sea back the marginal environments shift along with the shoreline and are superimposed on one another in accordance with Walther's law. In the rock record, remarkably similar sequences can be found serving as a record of the progradation of salt flats in a seaward direction over older intertidal and lagoonal sediments.

Submarine Slopes and Turbidites

One of the greatest advances in the study of sedimentary rocks took place in the middle of the twentieth century with the recognition that certain sedimentary rocks have been produced by turbidity currents. A turbidity current is a flow of dense, sediment-charged water moving down a slope under the influence of gravity.

Turbidity currents were first noticed in clear lakes, where flows were observed to form from muddy river water that hugged the take floor. These currents flowed for a considerable distance toward the center of the lake, slowing down and dropping their sediment only when they spread out and reached gentler slopes in the 1930s. Dutch geologist Philip Kuenen demonstrated in the laboratory that turbidity currents can attain great speed, especially when they are heavily laden with sediment and are moving down steep slopes. Sediment suspended in a turbidity currents behaves as part of the moving fluid and its presence increases the density of this fluid by as much as a factor of two.

When a turbidily current reaches a plane whose slope is more gentle than that from which it originated, it slows down and spreads out, dropping its sediment in the general sequence that we have now seen exemplified again and agains : First the coarse sediment falls from suspension, and then, much later, the fine material follows. The result is a graded bed that is often characterized by poorly sorted sand and granules at its base and by mud at the top. Such a graded bed is known as a turbidite.

A series of turbidity currents in the laboratory produced a succession of graded beds, leading Kuenen and others to recognize that similar turbidite successions are common in nature. This discovery prompted researchers to ask where in the oceans there might be slopes on which turbidity currents could be generated. Continental slopes were the obvious candidates. Sediments moving along these vast slopes would

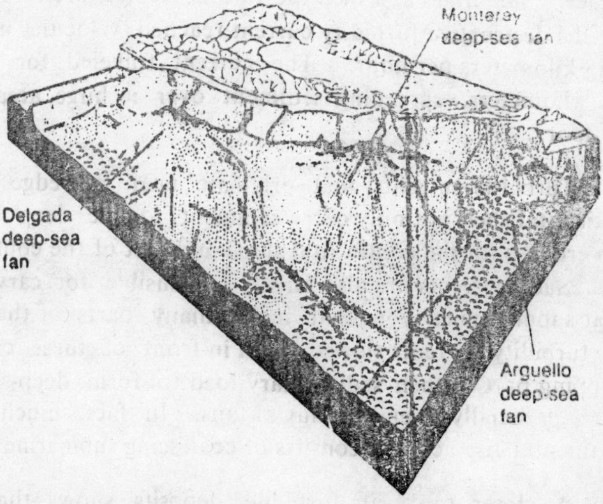

Fig. 5.6. Deep-sea fans spreading from submarine canyons along the coast of California.

be expected to be deposited in the deep sea, partly along continental rises and partly on abyssal plains.

The hypothesis that most large turbidity currents develop on continental slopes gained strong support from an event that was not fully analyzed until more than 20 years after it took place. In 1929, several transatlantic telegraph and telephone cables on the continental slope and rise off the Grand Banks of Newfoundland were mysteriously broken. An earthquake had taken place just before the cables were severed, and so it was assumed that a slumping of sediments resulting from the earthquake accounted for any breakage that had occurred close to the site of earth movements. Some cables, however, were snapped many minutes later and many kilometers downslope from the site of earth movement. When these events were evaluated more than 20 years after the fact, it was concluded that a turbidity current issuing from slumping sediment in the vicinity of the earthquake was the only agent capable to severing the deeper cables. Cores of sediment removed from the deepsea floor near these cables provided tangible evidence of this by revealing the presence of a turbidite at the sediment surface. The times at which the various cables broke revealed that the destructive turbidity current reached velocities up to 40 to 55 kilometers per hour. The current traveled for at least 700 kilometers, spreading sediment over a huge area of the sea floor.

Turbidity currents that originate near the edge of the continental shelf not only carry sediment to the deep sea but also erode both the continental slope and part of the continental rise. Such currents are also largely responsible for carving the great submarine canyons that incise many parts of the slope. The turbidity currents slowed down in front of these canyons, dropping part of their sedimentary load to form deep-sea fans that superficially resemble alluvial fans. In fact, much of the continental rise actually consists of coalescing submarine fans.

A closer took at turbidite deposits shows that their sands are typically graywackes that are quite unlike the clean sands of meandering-river deposits. This is only one of several ways in which a cyclical sequence of turbidites differs from the meandering-river cycle, although both show a grading of sediment from coarse to fine within each complete cycle. Another difference is that a turbidite sequence is usually only a few centimeters and seldom as much as a meter thick, whereas most complete meandering-stream cycles measure several meters from bottom to top. In addition, turbidites usually lack the cross-bedding that characterizes meandering-river channels.

Turbidites, in fact, exhibit a highly characteristic sequence of sedimentary structures. The base of a turbidite is often irregular because the earliest, most rapidly moving waters of the turbid flow have scoured depressions in the previous sedimentary surface. These scours are subsequenty filled in by the first sediments that settle as the current slows. When the base of a lithified turbidite is turned over for inspection, its irregularities, or "sole marks," often reveal the direction of water flow.

The characteristic sequence of sedimentary structures above the base of a turbidite is called a Bouma sequence, after the man who first recognized it. This sequence includes an upward decrease in grain size. Most turbidite beds, however, lack some features of the ideal cycle. Often, for example, the upper part of a cycle has been truncated by scour resulting from the arrival of the next turbidity current. The very top of a complete cycle usually consists of clays that accumulated slowly over a long period of time before the arrival of the subsequent turbidity current. This fine material might be described as representing "background" sedimentation that is periodically interrupted by violent turbid flows.

Pelagic Sediments

Turbidity currents and other bottom flows carry mud to the abyssal plain of the ocean well beyond the continental rise. Neither this source nor any other, however, contributes sediment to the deep sea at a high rate ; indeed sediment in most areas of the abyssal plain is accumulated at a rate of about 1 millimeter per 1C00 years : In the deep sea, sparse sediments from turbidity currents are joined by clays from two other sources. One source is the weathering of rocks produced by deep-sea volcanoes. Such clays are less abundant in the Atlantic than in the central Pacific, where volcanoes are common. Clays also reach the deep sea by setting from the pelagic realm, which they reach after travelling through the air as wind-blown dust or moving seaward from the land at very low concentrations in surface waters of the ocean.

Clays setting from the upper part of the water column are one type of pelagic sediment, Other types are contributed by small pelagic organisms whose skeletons become sedimentary particles when the organisms die. Whether clays predominate in any area of the deep sea depends on the extent to which they are diluted by the more rapid accumulation of biologically produced sediments. Some of the latter consist of calcium carbonate, while others consist of silica.

Calcium carbonate predominates in sea-floor sediment at low latitudes, where its fine grain size has led oceanographers to refer to it as "ooze," In fact, this calcareous sediment has often been termed "globigerina ooze," after the single-celled, amoeba like creature whose skeletons are common constituents. This label is usually not appropriate, however, because other genera of the same group (planktonic foraminifers) are often equally common or even dominant components of deep-sea calcareous ooze. Furthermore, skeleton sof altogether different kinds of organisms often predominate. The most important are coccoliths, armor like plates that surround coccolithophores, the single celled floating algae that are major components of tropical phytoplankton.

Calcareous oozes are abundant only at depths of less than about 4000 meters. The reason for this limitation is that calcium carbonate tends to dissolve as it settles through the water column. As pressure increases and temperature declines with depth, the concentration of carbon dioxide increases ; as a result, the water becomes undersaturated with respect to cal-

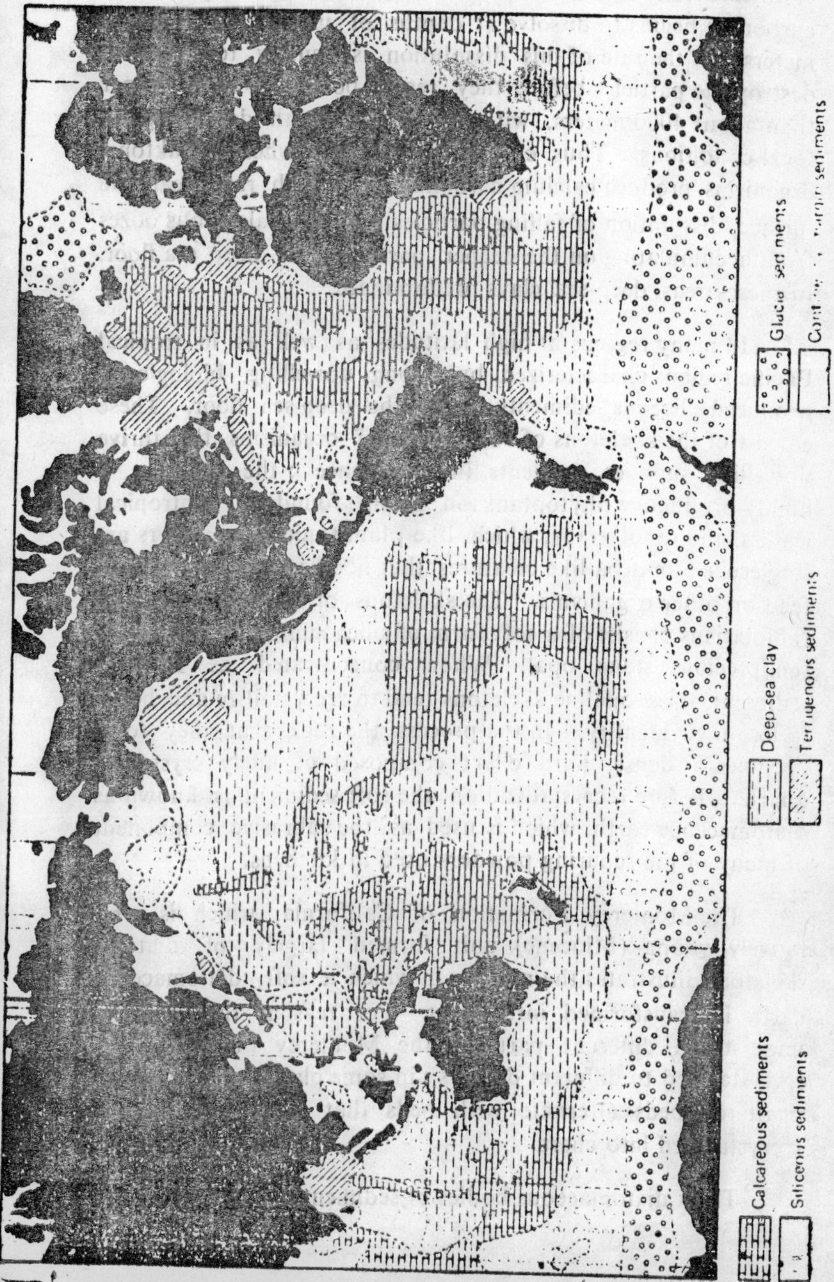

Fig. 5.7. The global pattern of deep-sea sediments.

cium carbonate and thus tends to dissolve it. Actually, calcium carbonate begins to dissolve at depths of just a few hundred meters, but the rate of this dissolution is usually too slow to destroy the particles before they reach bottom waters deeper than about 4000 meters, with the exact depth depending on focal conditions. Thus, even in the tropics, where planktonic organisms produce calcium carbonate at a high rate near the surface, dissolution of falling particles prevents calcareous oozes from accumulating on the deeper parts of the deep sea floor. In these areas, deep-sea clays predominate.

In many regions at high latitudes as well as in tropical Pacific regions characterized by strong upwelling, biologically produced siliceous sediments carpet the deep-sea floor. These consist of the skeletons of two groups of organisms that thrive upwelling supplies nutrients in abundance : the diatoms, a highly productive phytoplankton group found in nontropical waters, and radolarians, which, like planktonic foraminifers are singlecelled amoeba like creatures that float in the water and feed on other organisms. The skeletons of both diatoms and radiolarians consist of a soft form of silica similar to that of the semiprecious stone opal. When concentrated as siliceons sediment, these opaline skeletons tend to recrystallize and hence to lose their identity. In the process, the rock that they form becomes a dense, hard chart composed of finely crystalline quartz. Before recrystallization, the soft sediment is known as diatomaceous earth, which is used as the abrasive component of many of the scouring powders used in kitchens.

The Monterey Formation of California, which is a relatively young (Miocene) unit formed largely of diatoms skeletons, illustrates the manner in which a soft, diatomaceous earth is transformed into chert. For reasons that are not understood, different parts of the Monterey Formation are recrystallized to different degrees : in some places, soft, relatively unaltered beds alternate with beds that have been largely recrystallized into chert.

Thick diatomaceous bodies of sediment like the Monterey

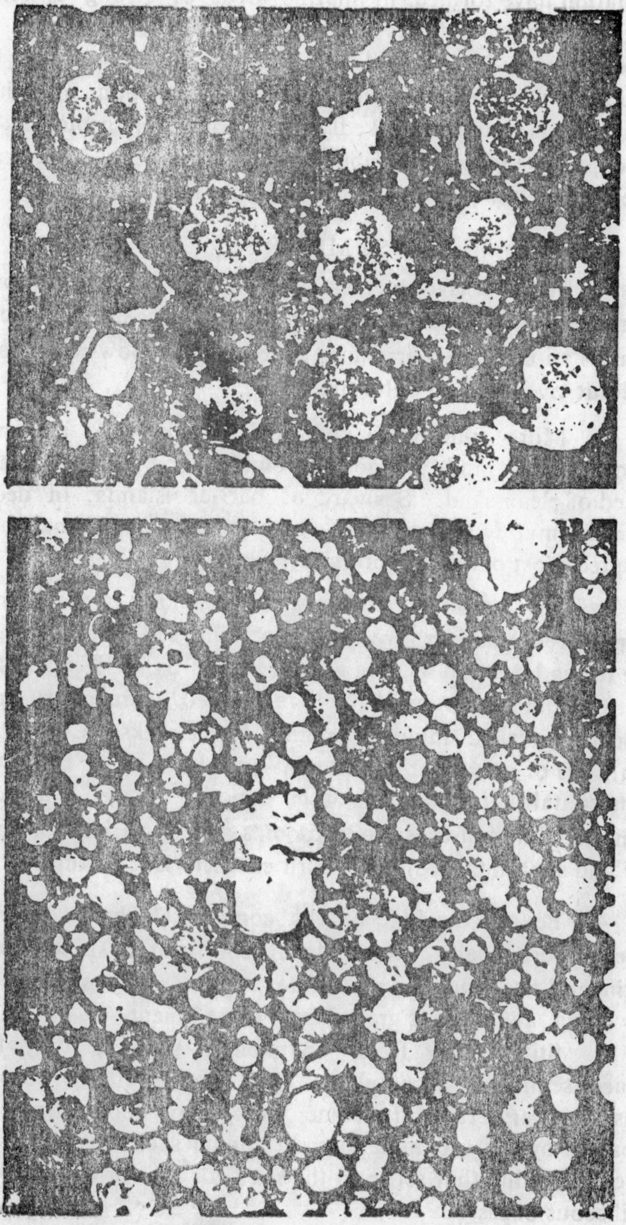

A B

Fig. 5.8. Foraminiferan (calcareous) oozes.

Formation have formed in marine areas of strong upwelling where diatom productivity has been usually high. Diatoms did not exist until late in Mesozoic time, however, and this illustrates an important point : The composition of pelagic sediments has changed markedly in the course of geologic tim as groups of sediment-contributing organism have waxed and waned within the pelagic realm of the ocean.

SUMMARY

1. Where a river meets a lake or an ocean, it drops its sedimentary loan to form a delta ; deltaic deposition typically produces an upward-coarsening sequence as shallow-water sands build out over deeper-water muds.

2. More widespread than deltas along the margin of the ocean are muddy lagoons bounded by barrier islands formed of clean sand. Seaward of barrier islands, in deeper water, are muddy sediments in which burrowing animals have disrupted most of the the original layering.

3. Coral reefs border shorelines in many tropical areas. A typical reef stands above the surrounding sea floor, growing close to sea level and leaving a quiet lagoon on its leeward side. Most reef limestones are largely unbedded sedimentary units supported by rigid internal organic frameworks. Coral reefs form parts of many carbonate platforms, although these platforms contain a number of other deposits as well (including submarine dunes of cross-bedded ooids and a variety of intertidal and supratidal sediments with characteristic features).

4. Beyond the edge of the continental shelf, turbidity currents periodically sweep down continental slopes to be continental rise and deep-sea floor, where they spread out, slow down, and deposit graded beds of sediment.

5. Still farther from continental shelves, only fine-grained sediments accumulate. Clay reaches these deep-sea areas at a very rate. In some areas, clay deposition is far surpassed by the accumulation of minute skeletons of planktonic marine life that settle to the sea floor to form calcareous or siliceous oozes.

6

Deep-Sea Environments

'Contrary to what is often imagined, biology is not a unified science. Marine biologiests will surely agree with Jacob (1974), but whatever their interests, whether physiological, biochemical, ethological, ecological or biogeographical, they know they must endeavour to be more than biologists Proper pursuance of their studies demands at least some familiarity with correlated knowledge about the physical nature of marine environments. As Pantin (1968) says, biological sciences are unrestricted' and their investigator must be prepared to follow their problems into any other science whatsoever.

'The cradle of life on earth is certainly the oceans' (Bernal, 1967), and throughout geological time sea water has been a benign medium for very diverse forms of plant and animal life. Since the beginning of Cambrian times, about 600 million years ago, Durham (1967) estimates, there have been at least four million and probably as many as ten million species of (preservable) marine organisms. In the modern ocean there are at least 160,000 species of animals, making them nearly twice as numerous as the non-insect forms of animal life on land. Moreover, certain forms, such as radiolarians, comb-jellyfishes, lamp-shells, arrow-worms, barnacles, cephalopods, beard-worms, echinoderms and tunicates live only in the sea. Foraminifera, sponges; cnidarians, bryozoans, sharks and rays are among those that are almost entirely marine.

Like Wordsworth's rocks, stones and trees, the ocean is 'rolled round in earth's diurnal course'. Even more than rocks and stones the ocean endures. 'The obvious things that no one comments on are often the most remarkable; one of them is the constancy of the total volume of water through the ages. The level of the sea, of course, has varied from time to time. During the ice ages, when much water was locked up in ice sheets on the continent, the level of the sea was lower than it is at present and the continental shelves of Europe and North America were laid bare. Often the sea has advanced over the coastal plains, but never has it covered all the land or even most of it. The mechanism of this equilibrium is unknown; it might have been expected that water would be expelled gradually from the interior of the earth and that the seas would grow steadily, or that water would be dissociated into hydrogen and oxygen in the upper atmosphere and that the hydrogen would escape, leading to a gradual drying up of the seas. These either do not happen or they balance each other' (Bullard, 1969).

Bullard writes above of the fully formed ocean that can be seen through the study of rocks and fossils. To trace the origin of all this water, or to appreciate why out planet might better be called Ocean rather than Earth, we must first realize that water, air and life (hydrosphere, atmosphere and biosphere are made of elements that are not prevalent in the earth's crust or even in volcanic rocks that have come from deep within the earth (Turekian, 1968). We are led to study the elements in the entire Solar System and driven to conclude that water was not derived by condensation during the formation of the earth but came from subsequent 'degassing' inside the newly formed earth. Relics of this process are apparent today in the explosions of volcanoes, for water seems to be part and parcel of molten rock (magma). The expansion of water vapour in the fierce heat below a plugged volcanic vent may eventually lead to an awesome explosion of steam, rocks and lava. It seems clear that all or most of the ocean grew from the 'volcanic breath' of the earth's mantle (Dineley, 1973). Indeed, the volcanic bed of the ocean extends over some seven-tenths of the earth's surface. We turn first to the bed then to waters.

The deep-sea floor

The configuration of the deep-sea floor is revealed by scrutiny of soundings, and nearly all have been made in the past fifty years. Since the nineteen-twenties, when echo-sounders came into use, a vast and varied, yet orderly topography has been brought into view. Survey of this submarine landscope with echo-sounders (and other sound sources) has also helped geologists to reveal the volcanic nature of the crust that bears the ocean.

There are three fundamental parts of the ocean floor: the continental margins, the ocean basin floors and the mid-oceanic ridge system. Over most of the earth the continental margin is dapped by a continental shelf, which is the submerged perimeter of the continent, ever subject to marine erosion and sedimentation. The shelf cants gently seaward (gradient about 1 : 1,0)0) and at depths ranging from 20 to 550 metres (average 133 m) gives way to the steeper gradient (>1 : 40) of the grandest escarpments on earth, the continental slopes.

The continental shelves, which are generally supported by sedimentary strata or by igneous and metamorphic rocks, have a mean width of 78 kilometres and underlie 75 per cent of the total expanse of the ocean. Excluding the littoral or intertidal waters, the seas over the shelves form the neritic province of the ocean : the rest is oceanic province and also the province of this book. But before leaving the shallow seas, which include the coral seas, it must be said that these waters hold a greater diversity of life and are much more productive than the deep ocean. For instance we know some 12,000 species of marine fishes and about 10,000 live in the shallow seas : the others are deep-sea species. Over nine-tenths of the world's annual catch of fishes (about 50 million tonnes) consists of species that depend on the most productive waters over the continental shelves. By comparison with the riches shelf waters, the deep ocean is a deserted living space.

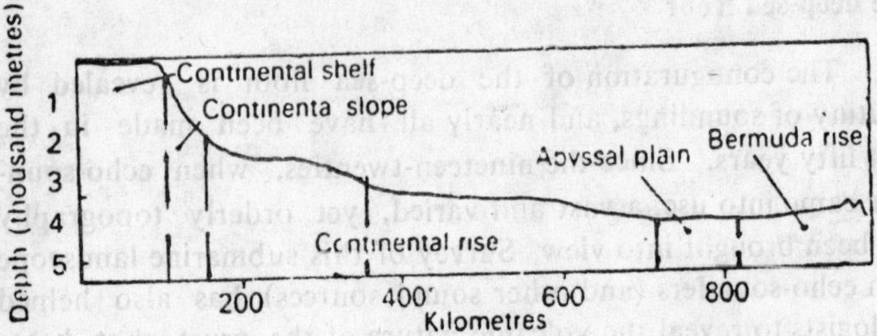

Fig. 6.1 Profile or the deep-sea floor off the northeastern United States. (Adapted from Heezen. Tharp and Ewing, 1959).

The continental slope descends to depths near 2,000 metres or more before merging with the gentler gradients (1 : 100-1 : 700) of the continental rise. Off the northeastern United States for instance the transition between continental shelf and slope is about 100 km off shore. At the foot of the continental slope, which is nearly 160 km off shore, the depth is nearly 2,000 metres. Towards the end of the continental rise, which is nearly 650 km from the shore, the depth is about 5,000 metress. This lowest reach of the continental margin averages about 4,000 metres in depth over the ocean. But over fringing areas of the western and northern Pacific Ocean, and off Chile and Peru, the continental margin descends into trenches with depths exceeding 6,000 metres.

Over most of the ocean the continental slopes and rises are covered with terrigenous or hemipelagic muds, mixtures of pelagic ooze and debris weathered from the continents and described as 'a sediment comprised of more than 30 per cent of continentally derived sand and silt grains; it is often green, black or slightly red, due to varying degrees of oxidation of the abundant organic matter'. More than 50 per cent of the particles forming this sediment are coarser than 30 microns in diameter, and the rate of sedimentation, which is appreciably

Fig. 6.2 General bathymetric chart of the ocean. (Adapted from Heezen and Hollister, 1971).

higher than that of any other deep-sea sediment, varies between depths of 5 and 100 centimetres per thousand years (Heezen and Hollister, 1971).

Nearness to the land and hence the relatively rich shelf waters has much to do with the high content of organic matter in terrigenous muds, which in turn goes far to being the basis for the relatively great productivity of bottom-dwelling life over and on the continental margins. In particular, the upper reaches of the continental slope are not only near to the land and shelf, but also comparatively near to sources of organic material from dead pelagic organisms. It is not surprising, then, that slopes and upper rises support more abundant populations than most of the deeper parts of the ocean.

Abyssal plains, typically the most level stretches of the deep-sea floor, are also covered and indeed formed by land-derived sediments. They lie at the foot of continental rises and are widespread in the Atlantic and Indian Oceans and very extensive off the continental margin of Antarctica. (Gradients range between 1 : 1,000 and 1 : 10,000.) The sediments over these plans resemble the sands, silts and gravels recovered from submarine canyons. Submarine canyons such as the Hudson Canyon, the Congo Canyon and the Ganges Canyon, fissure the continental slopes throughout the world. Most are only a few hundred metres deep but some cut 2,000 metres into the slope. All abyssal plains are connected by canyons or other channels to landward sources of sediment and down these fissures run fast turbidity currents charged with a great mass of sediment.

This sediment, which contains considerable amounts of organic matter, spreads over the plain. Photographs taken over the Sohm Abyssal Plain, south of the Great Banks of New-foundland '.. revealed very muddy bottom water, and it seems likely that some of the finest grained clays have not settled out from the most recent turbidity current' (Heezen and Hollister 1971). It is to be expected that some abyssal plains should be productive places for benthic life.

The ocean basin floors include the parts of the deep-sea floor between the continental margins and the major oceanic ridge systems. Apart from the deep-sea trenches the ocean basins form the deepest reaches of the ocean, mostly with depths of 4,000 metres or more. In the Atlantic Ocean, the main basins are the North American Basin (western North Atlantic), Canaries and Cape Verde Basins (eastern North Atlantic), Brazil and Argentine Basins (western South Atlantic), Angola, Cape and Agulhas Basins eastern South Atlantic) and the Atlantic-Antarctic Basin(southern South Atlantic). More than half the area of these basins extends to depths of 5,000 m etres or more, and much the same is true of the major basins of the Indian and Pacific Oceans. In the Northern Pacific Ocean, which is without a mid-oceanic ridge system, there is one vast North Pacific Basin, extending from the deep-sea trenches in the west to the continental margin of North America. Of course. in culinary (and relative) terms, the ocean basins look more like saucers than basins but their vast expanses are relieved here and there by islands and sea mounts, all the result of volcanic activity. 'Thousands of bountiful volcanic vents have poured out sufficient lava from the earth's mantle to build the massive underpinnings of oceanic islands, coral atolls and sea mounts. However, millions of abyssal hills, knolls and peaks have been produced by other less ample vents which failed to produce enough to build to the surface of the sea before their activity was snuffed out (Heezen and Hollister, 1971).

At depths of 5,000 metres or more the ocean basins are covered with the sediment known as red clay and at shallower reaches with calcareous oozes. Red clay, which covers half the deep-sea floor of the Pacific and a quarter of each of the Atlantic and Indian Oceans, is : 'A chocolate-brown, extremely fine-grained sediment ...It is composed of clay minerals, residue from dissolved plankton shells, wind-borne continentally derived silt, volcanic particles and in high latitudes red clay contains an admixture of ice-rafted pebbles, rocks and sand. Accumulation rates are the lowest of any deep-sea sediment, ranging from one-tenth to one millimetre per thousand years' (Heezen and Hollister, 1971). Indeed, red clay; which contain very little organic material, are the deserts of the deep-sea floor. Life on red clay consists largely of small invertebrates and are thinly spread.

The mid-ocean ridge system, which is a complex mountainous chain that winds round the world for a distance of 50,000 kilometres and forms a third of the deep-sea floor, is surely the most impressive major feature of our planet's physiography. Here is the volcanic birthplace of the deep-sea floor, which is formed of vast tectonic plates that grow slowly outward from the mid-ocean ridge. The crest of the ridge system '...is cleft by a deep median rift valley. Deep gashes cut into the sea floor nearly at right angles to the crest and flanks of the ridge continue across the ocean basin floor until they are obscured beneath the sediments accumulated in the continental margins. Sea mounts and islands lie in straight rows along these fracture zones and together with the linear scarps and ridges, which are parallel to the crest of the Mid-Ocean Ridge, they impose a vivid fabric-like texture on the ocean floor' (Heezen and Hollister, 1971).

The ridge is between 100 and 4,000 kilometres wide and from its foot-hills to the crest depths range from about 5,000 to 2,000 metres. The lower flanks are covered with red clay, the higher mountainous regions with calcareous oozes containing the remains of forminiferans and coccoliths. Actually, the valleys of the system are filled with ponds of sediment, which show clearly on the records of powerful, low-frequency echo-sounders. The peaks are just dusted with sediment. From ridge crest to ocean basin the covering of sediment gradually thickens.

Over its upper reaches, at least, there is presumably some variation in the rate of sedimentation, for the ridge system lies under waters of differing primary productivity (usually measured by the amount of carbon assimilated by the plant plankton under a square metre of the sea surface per day). Starting off east Greenland in the Jan Mayen area and proceeding through Iceland and along the mid-Atlantic to the latitude of Gibraltar, the ridge passes under waters of very high to high productivity.

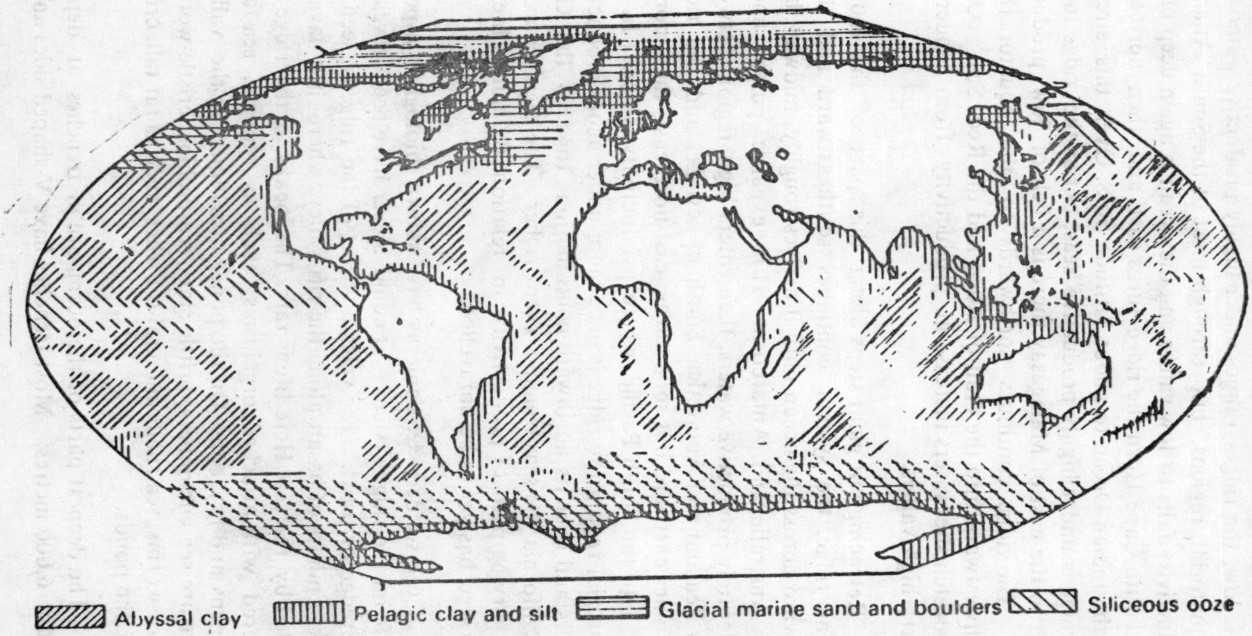

Fig. 6.3 Distribution of deep-sea sediments the white areas represent calcarous oozes. (Adapted from Heezen and Hollister, 1971).

Proceeding under the deep blue Sargasso Sea, where productivity is low, the ridges ranges under highly productive waters up to equatorial regions, but through most of the South Atlantic productivity falls to low values and the ocean is again deep blue. Level with Cape Horn the ridge turns east and then northeast into the Indian Ocean to about latitude 30°S. Over this stretch the ridge is under highly productive seas, but as the ridge turns north to the east of Malagasay there is a belt of low productivity. The ridge continues northwards to the equator then northwestward into the Gulf of Aden and the Red Sea. Over this stretch the waters increase in productivity from moderate to very high values.

Retracing out giant steps along this part of the Indian oceanic ridge, the system continues southeastward and then eastward across the ocean until it arrives roughly midway between Australia and Antarctica. This expanse passes under moderately productive waters, then enters high regions between New Zealand and Antarctica. South of new Zealand the ridge continues eastward but soon turns east-northeast then northward into the South Pacific, where productivity is low. In the equatorial region the ridge is once more under more productive waters and over its northward continuation towards the Gulf of California the seas become even richer The significance of this varying path of productivity in relation to animal life on the ridge has yet to be unravelled.

Even so, deep-sea cameras have shown biologists aspects of life on the ridge system that could never have been revealed by dredge or trawl. The volcanic heart of the ridge is seen in broken pillars of basalt along the rift valley, where hot lava is chilled by the sea. Here life is rare. The peaks of the ridge are covered with attached organisms, such as sponges, sea-fans, sea-pens and bryzoans. On the ponds of sediment in the valleys there are sea-cucumbers, brittle-stars, sea-urchins, bristle-worms, acorn-worms, sea snails and crabs. Fishes, such as rat-tall, cruise over the ponds.

The deepest parts of the ocean are in trenches at depths below 6,000 metres. Most trenches have V-shaped sides and a level floor. Thirty-one have been listed, and most lie along the margins of the Pacific Ocean : others indent the borders of the northern Indian Ocean and in the Atlantic follow the outer loops of the Caribbean and Scotia Arcs. Deepsea trenches are elongated and narrow depressions with lengths ranging from a few hundred to over 2,000 kilometres and mean widths from 25 to 120 kilometres.

Starting from New Zealand and moving clockwise round the Pacific, the main trenches and their maximum depths (in metres in parentheses) are : the Kermadec Trench (10,047), Tonga Trench (10,800), New Hebrides Trench (9,165), Philippine Trench (10,497) Mariana Trench (11,524), Ryukyu Trench (7,507), Idzu-Bonin Trench (9,810), Japan Trench (8,412), Kurile-Kamchatka Trench (10,542), Aleutian Trench (7,679), Middle American. Trench (6,662) and the Peru-Chile Trench (8,055). The main trenches of the Indian Ocean are the Chagos Trench (5,408), Java Trench (7,450), Amirante Trench (9,074), Mauritus Trench (5,564) and the Diamantina Trench (8,230). In the Atlantic lie the Puerto Rico Trench (8,385), Cayman Trench (7,930), Dominican Trench (ca. 6,200), Romanche Trench (7,856) and South Sandwich Trench (8,428). Most trenches, then, are parallel to island arcs of parts of the continents with coastal cordillera. It is now evident that such trenches are formed along collision lines between tectonic plates.

Though the deepest parts of the ocean, trenches are not necessarily the most barren parts. The deepest levels are filled with land-derived sediments, which may be rich in organic material and several trenches underlie waters of high productivity. Invertebrate life may thus be relatively abundant, and indeed, diverse. Bruun called these animals the hadal fauna, for present exploration indicates that each trench contains a proportion of (endemic) species that have yet to be found elsewhere.

Oceanic waters : the pelagic environment

Water, salt and heat balance

Physical oceanographers study water, salt and heat balance

in the ocean, and so do physiologists in living organisms. Both ocean and organisms tend to maintain steady states. One that is often taken for granted, the principle of conservation of volume, follows from the fact that water is no more than slightly compressible. Thus, the fact that water flows out of basins such as the Mediterranean and Red Sea with no change in sea level means that water must be flowing in at the same rate.

Elsewhere the conservation of volume has profound effects on the productivity of the ocean. For instance, in the Antarctic Ocean and off parts of the western sea board of the African and American continents, surface waters are displaced and water rich in nutrient salts moves up to fill their place. We shall see later that the upwelling leads to prolific crops of plant plankton.

The sea gains water from the land, rain and melting ice and loses it through evaporation and freezing. Most of the earth's rain water comes from the very surface of the ocean (carrying much particulate matter) and much finds its way back to its source. Clearly, salinity increased by evaporation and reduced by precipitation. Indeed, if the mean values of surface salinity and the difference between evaporation and precipitation (in cm) are plotted against latitude, there is a close parallel between the two curves. Salinity, which is the total amount of dissolved material in parts per thousand (%) of sea water, is at a minimum (ca, 34.5%) near the Equator, rises to a maximum (35.8) in about latitudes 20°N and 20°S and then decreases to 34.0% or less towards high latitudes.

Though the salinity varies, the proportion of the main ions in sea water is almost constant, which C.R. Dittmar discovered by making careful analyses of 77 water samples collected all over the world by HMS *Challenger*. Thus, percentage values are ; chloride 55-26 ; sulphate 7-68 ; magnesium 0.76 ; calcium and strontium 1.20 ; potassium 1.12 ; and sodium 30.58. This uniformity of relative composition is due to constant mixing and circulation and it is vital to marine plants and animals which all have near-oceanic cell sap or body fluids. If any one of the main ions changed in proportion from place to place, marine organisms would need to be ever using energy reserves to maintain a steady inner state. Even to, some animals do contain concentrations of certain ions that differ considerably from those in sea water. For instance, relative to sea water, magnesium ions are lower in concentration in the body fluids of many crustaceans and chordates but higher in some bivalves, gastropods and cephalopods. Potassium ions, which are particularly vital for nervous and muscular systems, are concentrated by molluscs and crustaceans. In the body fluid of the squid, *Loligo*, the concentration of potassium is more than twice that of sea water.

There is also the principle of conservation of salt, which '...asserts that the total amount of dissolved salts in the ocean is constant. When one first learns that the rivers of the world centribute to the sea a total of about 3×10^{12} kg of dissolved solids per year, the conservation of salt seems to be contradicted. In principle it is, but in practice it is contradicted only to a negligible extent. The amount of salt in the oceans is about 5×10^{19} kg and therefore the amount brought in each year by rivers increases the ocean salinity by about one part in three million per year. But we can only measure the salinity of sea water to an accuracy of about 20.003‰, or about 300 parts in three million taking the mean ocean salinity as 35‰. In other words, the oceans are increasing in salinity each year by an amout which is three hundred times smaller than our best accuracy of measurement. So fall all practical purposes we can assume that the average salinity of the oceans is constant at least over periods of tens or even hundreds of years' (Pickard, 1963).

The surface of the sea covers 71 per cent of the earth and thus intercepts most of the sun's heat that falls on our planet. Even so, most of the ocean is cold, for though tropical latitudes gain much heat, cold waters, especially in polar regions, sink and fill the ocean basins.

Radiant energy from the sun reaches the earth directly or as radiation scattered by the atmosphere, particularly by clouds (sky radiation). Atthe sea surface, some of this energy is scattered : the rest enters the sea, is absorbed and raises the temperature of the water. At already intimated, the intensity of solar energy falling on a unit area of surface from a beam of given cross-section decreases from equator to poles (as the cosine of the latitude). At a given latitude the diurnal and seasonal range in the sun's elevation will have a differential radiant effect like that of latitudinal variation. Again, when the sun is directly overhead (90 elevation) a unit area of sea surface will receive more radiant heat from a given beam than at slanting elevations (and the amount decreases as the size of the angle of elevation). Concerning the reflection of direct radiant energy at the sea surface, the lower the elevation of the sun, and the more disturbed the water surface, the greater the amount of reflection. For sky radiation, which comes from all directions and varies mainly with cloud cover, about a tenth is reflected.

The sea surface loses heat through long-wave radiation to the atmosphere, through heat lost by evaporation and through heat conducted away from the surface, either into the water or out of the water. If these three loss quantities are Qb, Qe and Qh respectively and Qs is the rate of flow of solar energy through the sea surface, then the heat balance is

$$Qs = Qb + Qe + Qh$$

The outstanding thermal feature of the ocean, as already hinted, is that there is an upper-warm layer of water, mainly in tropical and subtropical latitudes, floating, as it were, on a cold ocean. Anton Bruun, who believed the isotherm for 10°C to have wide significance in marine biogeography, called the warm water above this isotherm the thermosphere and the cold underlying waters the psychrosphere. In the Atlantic and Pacific Oceans the thermosphere extends between latitudes 50°N and 50°S and to depths between 300 and 1,000 metres. If the warm and cold water spheres are so defined, the former would seem to occupy no more than one-tenth of the entire volume of the ocean. Indeed, three-quarters of this volume has temperatures between 0° and 6°C (and salinities between 34.6% and 34.8%). The mean temperature of the entire ocean 3.5°C (and the mean salinity 34.7%).

The temperature structure of the ocean has, of course a more complex pattern. A profile of temperature through the

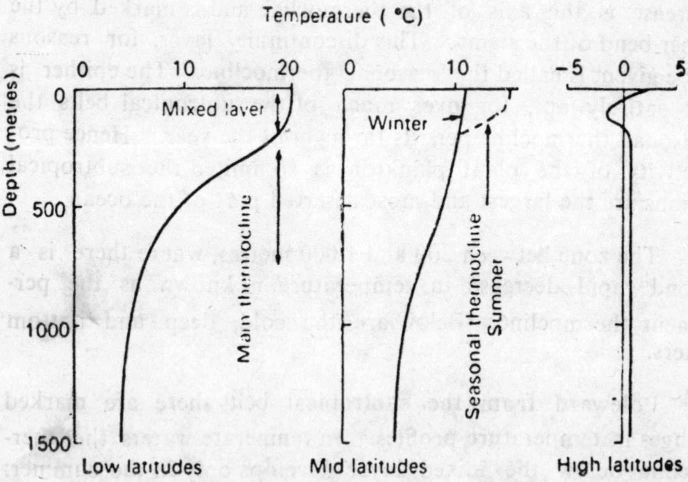

Fig. 6.4 Typical mean temperature profiles in tropical, temperate and cold latitudes of the open ocean, (From Pickard, 1963)

warm and cold water spheres traces a sigmoid curve. The upper most, vertical part of the curve shows an isothermal surface layer ranging from about 10 to 200 metres in thickness. Temperatures in this layer range between 30° and 15°C, depending on season and latitude. In a zone below this, extending between depths of 300 and 1,000 metres, temperatures fall rapidly to values near 5°C, then in the rest of the water column the temperature decreases slowly to levels between 1° and 2°C at 4,000 metres, the mean depth of the ocean.

The surface layer is mixed to isothermal consistency by the wind, and is thus called the 'mixed layer'. It becomes thicker in winter, especially in middle latitudes In the western North Atlantic, for instance, the mean depth of the mixed layer is between 40 and 120 metres in winter and between 10 and 60 metres in summer. The layer is separated from cooler underlying waters by a thermocline, a layer in which there is a rapid decrease of temperature with depth. The depth of maximum decrease is the axis of the thermocline and is marked by the upper bend of the sigma. This discontinuity layer, for reasons to be given, is called the 'seasonal' thermocline. The epither is not entirely apt, for over much of the subtropical belts the 'seasonal' thermocline persists throughout the year. Hence productivity of the plant plankton is so limited the subtropical regions are the largest and most deserted part of the ocean.

The zone between 300 and 1,000 metres, where there is a second rapid decrease in temperature, is known as the permanent thermocline. Below are the cold, deep and bottom waters.

Poleward from the subtropical belts there are marked changes in temperature profiles. In temperate waters the thermocline below the mixed layer develops only in the summer: hence the term 'seasonal'. The permanent thermocline is also much less marked, and one reaching polar waters, thermoclines have vanished. Indeed, there is little change in temperature from to bottom. In the coldest polar waters there is an increase from negative values (—1.9°C to 0°C) to slightly values at depths below 200 metres.

The ocean is immense and water absorbs great quantities of heat with relatively little change in temperature. Except for liquid ammonia, the heat capacity of water is the highest of all solids and liquids. Thus, after leaving the surface layer, great quantities of heat are moved by the circulation from one part of the ocean to another. (The transfer of water properties through currents is termed 'advection', to distinguish it from transfer through di fusion.)

Moreover, there are extreme ranges or sudden changes in temperature in the ocean, which in this respect (and others) is a more benign environment than the atmosphere. On land the most abundant and ubiquitous vertebrates are birds and mammals, which are warm-blooded and able to regulate their body temperature. In the sea the warm-blooded homeothermal whales seals and penguins, which had land-dwelling ancestors, range from polar to tropical regions. The only inveterate warm-blooded dwellers of the ocean among the vertebrates are members of the tuna family (Thunnidae) and makoshark family (Isuridae). The bluefin tuna, which has some power of temperature regulation. ranges from warm waters off the Bahamas to cool sea off Norway.

Even so, many marine organisms are eurythermal, able to exist over a wide range of temperature. The most remakable are the diverse invertebrates and fishes that migrate up and down each day between cool deep waters (near 10°C) to surface waters in the tropics with temperatures in excess of 20°C. There are planktonic organisms that live in the ocean from polar to tropical regions.

But any one of the these species may consist of temperature races, each adapted to a relatively narrow range of temperature, the entire complex covering the span of environmental temperatures. Temperature tolerances of marine organisms range from the seeurythermal (generalist) kinds to specialists that exist within narrow limits of temperature and are described as stenothermal.

Density

Density in the ocean is fundamental to marine science. For physical oceanographers, a synoptic survey of density helps them to discover the origins, equilibrium levels and spread of water masses. The density of marine organisms matches or is a little more than that of the sea, and biologists are thus led, inter alia, to investigate the buoyancy relations of diverse plants and animals. For animals, a crucial metabolic advantage of weightlessness in winter is that they can stay at a preferred level with the minimum expenditure of energy.

Density which is expressed in grams per cubic centimetre, depends on pressure as well as on salinity and temperature. Pressure in water increases by one atmosphere for each increase in depth of to metres, and water is slightly compressible. Thus at the surface, sea water of salinity 35% and temperature 5·0°C has a density of 1.012813, which at a depth of 4,000 meteres and hydrostatic pressure of 400 atmospheres is increased by compression to a value of 1.04849.

Since physical oceanographers usually compare water masses at the same depth or over the same interval of depth, densities are reduced to atmospheric (i,e. hydrostatic pressure, $p=0$), but salinities and temperatures are taken as measured at the depths involved. In the open ocean values of density range from about 1.024 to 1.030g/cm^3. Naturally, the saltier the water, the more its density, while warm water is less dense than cold. More precisely, the rate of increase of density with salinity at atmospheric pressure is about 8×10^{-4} gm/cm^3 per 1%. The rate of decrease of density with increasing temperature is more involved and ranges from 5×10^{-5} per 1°C change at 0°C to 34×10^{-6} per 1°C change at 30°C.

The ocean is more than 800 times denser than the atmosphere, and is thus more buoyant by the same factor. Sea water sustains, in more than one sense, a marvellous diversity of organisms from bacteria to blue whales. It is hardly fortuitous that 'protoplasm', which arose in the sea, is only a little denser than its birth fluid, or that the cell sap of plants and the body fluids of animals have a near-oceanic salt balance. Marine animals almost entirely depend, directly or indirectly, on the microscopic plants in the plankton, which have buoyant and motile means to stay in surface waters, where light is strong enough for photosynthesis. Beside supporting whales, seals and fishes, all with firm muscular bodies and heavy skeletons, the buoyant properties of the sea have led to its exploitation by forms that consist largely of a fragile, gelatious skeleton. Medusae jellyfishes, siphonophores and pelagic tunicates are among such forms.

Since marine animals have much the same compressibility as the sea, their change of density with depth will match that of their environment. Gases, of course are highly compressible, and animals that depend on a gas-filled float for neutral buoyancy have evolved means of regulating their density against changes of pressure.

Changes of pressure have less obvious effects on organisms. Certain fishes and planktonic animals of coastal waters respond in various ways to quite small changes in pressure, and it will be interesting to discover if the same is true of their relatives in the deep sea. Changes of pressure have effects on the activities and enzyme systems of marine animals. Certain deep-sea species tolerate wide ranges of pressure; others seem to be less tolerant.

Water masses and circulation

Through tracing the origin and distribution of each oceanic water mass, physical oceanographers have helped marine biologist to understand better the distribution of deep-sea organisms. Water masses keep their identity and support certain characteristic communities of invertebrates and fishes.

Water masses have individual ranges of temperature, salinity and density, and in 1918 the Norwegian oceanographer, Bjorn Helland-Hansen, realized how these properties could be used effectively to define a water mass. Besides plotting the values of temperature and salinity against the depths of sampling at any particular station, Helland-Hansen found that characteristic thermohaline curves are produced when one property is plotted against the other. Such as curve is known as a TS diagram, and there is an advantage in using temperature, salinity (and density) as identification marks of water masses. Near the surface, as we saw, these properties may be changed by variations in solar radiation, evaporation, precipitation and so forth but away from the surface they are unchanged, except through mixing processes. The are 'conservative properties'. Eeven so, other features, such as oxygen content or concentration of nutrients (e.g. phosphates), can help in identifying water masses. When using these fearures one has to remember that they are 'non-conservative' properties: their concentrations are changed by living organisms.

If there are enough stations in depth and area for synoptic survey, one can see where a water mass begins and ends. That each water mass has a characteristic TS diagram may be appreciated when the TS curves for an ocean are plotted on one graph. Wider understanding comes from comparing the oceans, which have certain water masses in common. Thus, the Atlantic Ocean, India Ocean and South Pacific Oceans contain water masses that are given these names : Antarctic Bottom Water, Antarctic Intermediate Water, Subantarctic Water and Central Water. Apart from the last kind, these water masses, as implied by their names, are formed in antarctic and subantarctic regions of the Southern Ocean. Their origin is near the surface and, together with water masses formed near the surface in high latitudes of the North Atlantic and North Pacific, they fill about three-quarters of the ocean, which as we saw, is predominantly cold.

We begin then, in the Southern Ocean, which merges widely with the three other oceans. After moving into the Atlantic Ocean, which has been well explored, we will then compare and contrast the Atlantic with the Pacific and Indian Oceans.

During the winter the surface waters around Antarctica become colder and saltier, and by summer, when the days are longer and ice melts, become warmer and less saline. In winter surface temperatures are between−1.9°C and 1°C, rising to between—1°C and 4°C in summer. Winter conditions are particulary severe over the shelf of the Weddell Sea, where temperatures reach—1.9°C and the formation of sea ice leads the water of relatively high salinity (34.6%). Such water has a high density (I-0279), and after mixing with Circumpolar Water, it sinks down the continental slope and flows away from Antrarctica over deeper reaches of the ocean floor. This is Antarctic Bottom Water and it moves northward and eastward into the basins of the Southern Atlantic, Indian and Pacific Oceans. In the Atlantic this water type has been traced as far north is 40°N, about the latitude of Spain. There is evidence of Subpolar Bottom Water off eastern Greenland and in the Norwegian Sea, but this water clearly nothing like the volume of Antarctic Bottom Water.

North of a narrow east wind drift near the land, surface waters off Antarctica are driven clockwise round the continent as a broad west wind drift. Due to the force inherent in the earth's rotation (Coriolis Force), there is a northerly component in this drift. Over southerly reaches, the properties of Antarctic Surface Water, which has a thickness of 100 to 250 metres, vary

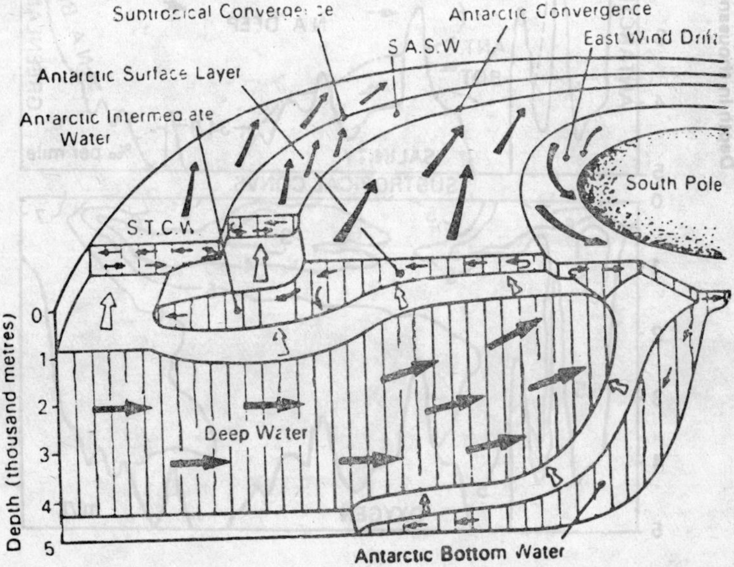

Fig. 6.5. Three dimesional represen'ation of the water masse. and circulation of the Southern Ocean.

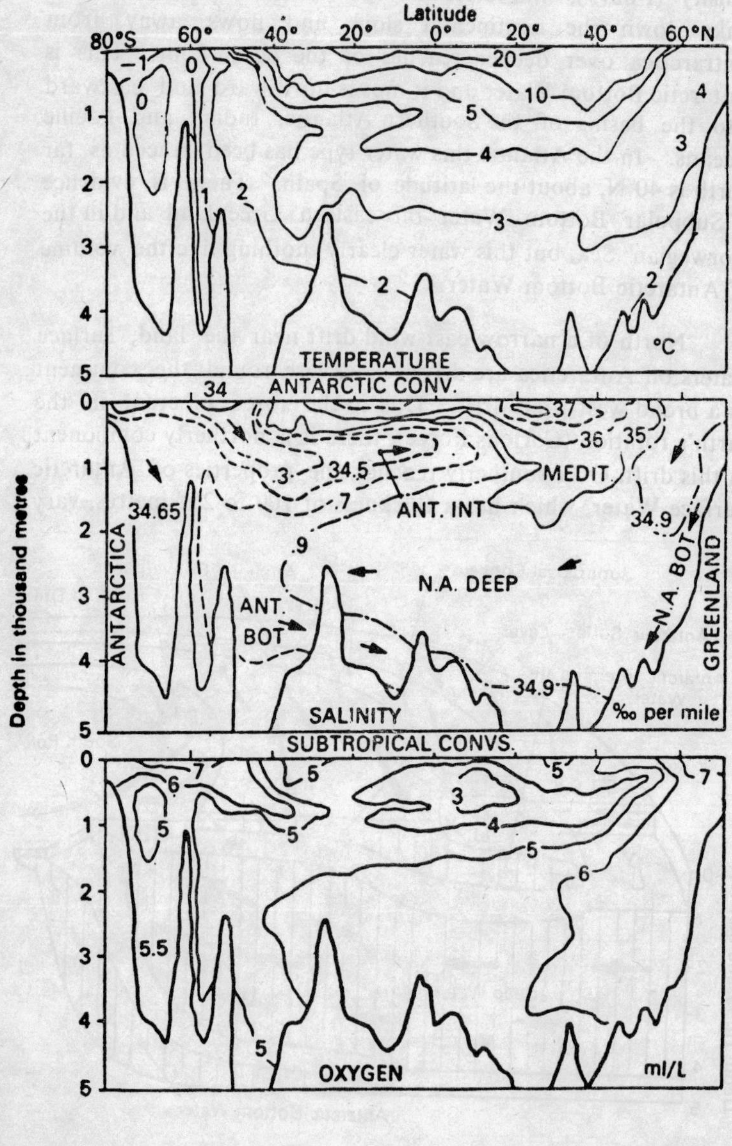

Fig. 6.6 South to north longirudinal section of water characteristics along the western trough of the Atlantic Ocean. (From Pickard, 1963, after Wust)

with the seasons (winter temperatures reach—1.9°C and salinities nearly 34.5%), but during their oblique drift to the north these waters absorb summer heat and at latitudes between 60″S and 50°S temperatures are close to 2°C and salinities below 34%. Between these latitudes and along a narrow circumpolar front called the Antarctic Convergence, Antarctic Surface Water of higher temperature (4°C or more) and begins to sink. On the way down it mixes strongly with deeper and warmer waters that are rising above Antarctic Bottom Water and after finding its level at about 1,000 metres, continues north as Antarctic Intermediate Water (temperature 2°C, salinity 33.8%). In the Atlantic this water mass crosses the Equator in the west. North of the Equator in the east a smaller intermediate water mass is formed by the sinking of Mediterranean water to 1,000 metres or more as it flows into the Atlantic. The warm deep water is not formed in the Southern Ocean but far to the north in the surface waters of the North Atlantic. South of Greedland at about latitude 60°N relatively saline waters that are chilled in winter, sink and then flow south between depths of 1,500 and 4,000 metres as the very voluminous North Atlantic Deep Water Mass.

North of the Antarctic Convergence at about latitude 40$ there is a second circumpolar front, the Subtropical Convergence, where cool Subantarctic Water meets a warmer mass of water. Here, over a much broader front than met at the Antarctic Convergence, water sinks to levels between Antarctic Intermediate Water, and the surface mixed layer to form South Atlantic Central Water. North Atlantic Central Water is formed along another broad front at about 40 N in the northwestern Atlantic. These Atlantic Central Waters, which extend to depths of 300 metres on either side the Equator and deepen to 600-900 metres at mid-latitude, have temperature ranging from about 17 to 7°C and salinities from 36.5 to 34.5%. It will be seen that the Central Water Masses form most of the thermosphere.

The Southern Subtropical Convergence may be taken also

as the southern boundary of the Atlantic, Indian and Pacific Oceans. The rest of the ocean is formed of 'seas'. One group, such as the Arctic Sea, the Bering Sea, the Sea of Okhotsk, the Gulf of Mexico, the Caribbean Sea, the Mediterranean Sea the Red Sea, the Sea of Japan and the South China Sea, are partly enclosed by land or island arcs, and thus have restricted openings into the main ocean. The term 'sea' is given also to certain unconfined parts of the ocean that have individual oceanographic features. The Barents Sea, the Norwegian Sea,, the Arabian Sea and the Tasman Sea are among such seas.

Conditions in the Pacific Ocean and Indian Ocean may now be seen in comparison and contrast to the above outline of water masses in the Southern Ocean and Atlantic Ocean, Antarctic Bottom Water and Antarctic Intermediate Water move into southerly parts of the Pacific and Indian Oceans. Intermediate Water in both oceans flows north to meet Equatorial Water Masses, soon to be described. Intermediate Water informed also in the northern Pacific, evidently by vertical mixing in the subantarctic region, not by the sinking of surface water. In the northwestern Indian Ocean an intermediate water mass between depths of 1,000 and 1,500 metres may be traced to the sinking of warm and saline waters (salinity 36%, temperature 15°C) that flow out of the Red Sea.

There is little or no formation of deep water in the Pacific Ocean, and from a level of 2,500 metres downwards the Deep and Bottom Water (temperature 1° to 3°C, salinity 34.6 to 34 8%) must come entirely from the Antarctic Circumpolar Current. This is also the source of some of the deep water in the Indian Ocean ; as in North Atlantic Deep Water, which enters at depths below 2,000 metree in the southwest. Two more deep-water masses are formed within the ocean itself. North Indian Deep Water forms in the Arabian Sea from Red Sea water and a much smaller outflow from the Gulf of Oman. The Red Sea Water moves towards Africa and may betraced as far south as 5° to 6° S. Further south, between 10° and 16°S, North Indian Deep Water mixes with Antarctic Intermediate Water and Antarctic Bottom Water and is transformed into

South Indian Deep Water (temperatures 3.0 to 1.0°C, salinities 34.78 to 34.60%).

The Central water Masses in the South Pacific and South India Ocean arise, of course, at the Southern Subtropical Convergence. In the North Pacific Central Water is formed along a front at about latitude 30°N. North of this is a broad ocean-spanning mass of Subarctic Water of low salinity (33.5 34.5%) and rather low temperatures (2 to 4°C). In both oceans the Central Water Masses lie between depths of about 100 and, 1,000 metres and have temperatures ranging from 17 to 4°C, salinities from 36.3% to 34.2 %.

Between the Central Water Masses of the Pacific and north of the Central Water in the Indian Ocean is a vast Equatorial Water Mass. There is no counterpart in the Atlantic Ocean. Pacific Equatorial Water extends right across the ocean, decreasing in width towards west and reaching a depth about 800 metres. It has temperatures of about 15° to 5°C, salinities of about (35.3 to 34.7%). Indian Equatorial Water which has a small range of saiinity (35.2 to 34.9%), extends across the ocean to the north of latitude 10°S and reaches a depth of about 1,000 metres, Indian Central Water merges with Antartic Intermediate Water at a depth of some 1,000 metres. In the pacific and Indian Ocean, Equatorial and Central Waters form most of the thermosphere. The uppermost part consists of the surface mixed layer.

This ends a brief survey of the oceanic water masses seen in terms of temperature, salinity and density (based largely on Pickard, 1963). Water masses maintain their thermohaline (and other) features, which was well seen when observations made in the Atlantic during 1925 to 1927 by the German *Meteor* Expedition were compared with those made during the International Geophysical Year (IGY, 1957-58) on the ships of the Woods Hole Oceanographic Institution USA and the National Institute of Oceanography (UK). Relevant comparisons of the two sets of observations showed that the distributions of temperature, salinity and dissolved oxygen were almost identical. Marine biologists who know a particular area through repeated observations find also that communities of deep sea animals may maintain their identity. Diverse species seem to bound to the physical and biological conditions of water masses : others transcend these conditions.

After thirty years, it seems astonishing that values of dissolved oxygen were still the same in certain part of the Atlantic. Through a given 'volume' of the ocean, the circulation moves at a certain rate and carries a certain quantity of dissolved oxygen, which is used by resident organisms in the processes of life and decay, Does the correspondence between the Meteor and IGY measurements imply that in some parts of the ocean there is a steady state in the quantity of dissolved oxygen ? Is the supply of oxygen so great compared to the oxygen requirements of the organism that methods of oxygen analysis are unable to detect the withdrawal of oxygen by life ? At mid-depths in some parts of the ocean the oxygen content is minute but life abundant. Life in these oxygen minimum layers will be considered later.

Convergences, where surface waters rich in oxygen sink into the depths, have been called the lungs of the ocean. The atmosphere is the main source of oxygen dissolved in sea water. The plants in the plankton also produce oxygen during photosynthesis, and at times the surface layer may be supersaturated with this gas. The oxygen content of water is expressed as the number of millilitres (ml) of oxygen gas dissolved in one litre (l) of seawater at NTP. Values range from 0-8 ml/l.

Down from the convergences the circulation of the water masses carries oxygen to nearly all parts of the ocean below the surface layer. The exceptions are the Black Sea and Cariaco Trench (in the Caribbean Sea) where subsurface layers are devoid of oxygen and carry hydrogen sulphide, sure signs of stagnant conditions. As a water mass sinks and finds its level in the ocean, the contained oxygen is used by living organisms and in the oxidation of organic detritus. Once it has left the surface a water mass can only gain oxygen by mixing with another water mass of higher oxygen content. If the circulation is sluggish a water mass will lose much oxygen to its life after travelling several thousand miles. But how fast is sluggish ?

Natural radiocarbon (C^{14}) is the most useful of the radionuclide tracers of oceanic circulation. The sea surface gains radiocarbon from the atmosphere. Thus, the concentration of C^{14} in part of a water mass will depend on how long this part has been below the surface (the residence time) and on the rate of gain or loss of C^{14} to local water masses. To use this method, then, one must make certain assumptions about the exchange of C^{44} between the atmosphere and ocean and about mixing between water masses. In a survey of the distributions of C^{14} in the Pacific and Indian Oceans much emerged. Deep waters were less active than those near the surface, showing, as one would expect, their greater age. In the Pacific, for instance, the activity of the deep water decreased from south to north, which reflects the ageing of the water as it moves north. The northerly component in the deep circulation was calculated to have an average rate of 0 05 cm/sec (43 meters per day). Since the deep water moves so slowly *and* originated in antartic region, it is hardly surprising that values of oxygen content at levels of 2,000 metres or more have fallen from about 4.5 ml/l in the South Pacific to between 0.5 and 3 ml/l in the North Pacific, Indeed, in the eastern Pacific between 10 and 20 N there is a pronounced oxygen minimum layer with values down to less than 0.1ml/l between 200 and 1,000 metres. Here productivity is high and it seems that the requirements of life, combined with slow circulation, are the major causes for the oxygen minimum layer.

As already implied, the distribution of radiocarbon can be used to calculate mean residence times for the various water masses. For deep Pacific water the residence time is about 1,000 to 1,600 years, which is approximately twice that of deep water in the Atlantic and Indian Oceans. In Atlantic deep-water values of dissolved oxygen (3 to 6.5 ml/l) are higher than these in Pacific deep water, which seems reasonable in view of the less sluggish circulation of the former. But deep currents are not invariably slow. At all events, the slow pace of the deep waters may be appreciated in contrast to surface waters, which have a residence time of 10 to 23 years and a lively circulation.

Patterns of circulation near the surface

Circulation in the upper waters is turned by part of the sun's energy that falls on the sea. Evaporation at the surface, which may be facilitated by the stirring motions of microscopic plants and animals in the surface film, yields potential energy in the form of water vapour. Minute particles of salt and organic material also leave the surface film, electedit seems by breaking bubbles (MacIntyre, 1974), Water vapour and particles are swirled into the atmosphere, where the particles form the nuclei for the condensation of water vapour which releases latents heat, the main source of energy for atmospheric circulation. Winds that blow over the sea raise waves and drive ocean currents. The wind-driven circulation of the ocean thus differs from the thermohaline circulation of the water masses, which as we saw, is maintained by the pull of gravity on dense water types that sink and then flow along stratified levels of equilibrium. The wind-driven circulation consists essentially of currents that flow horizontally in the upper few hundred metres of the ocean. Due to the force inherent in the earth's rotation (Coriolis Force), these currents move along lines to the right of the wind direction in the northern hemisphere and to the left in the southern.

The patterns of wind-driven circulation in the three main oceans are broadly similar, as one might expect. In equatorial regions the Trade Winds drive a current system that A. Defant called 'the backbone of the circulation'. Near or at the equator there is a westerly flowing South Equatorial Current that it kept in motion by the southeast Trades. To the north of the equator a parallel, westerly flowing North Equatorial Current is driven by the northeast Trades. Both currents cross the ocean and pile up water that returns between them as an Equatorial Counter Current. This current spans the entire width of the Pacific Ocean but is prominent only over the eastern half of the Atlantic. There is a parallel, triple current system in the equatorial Indian Ocean from November

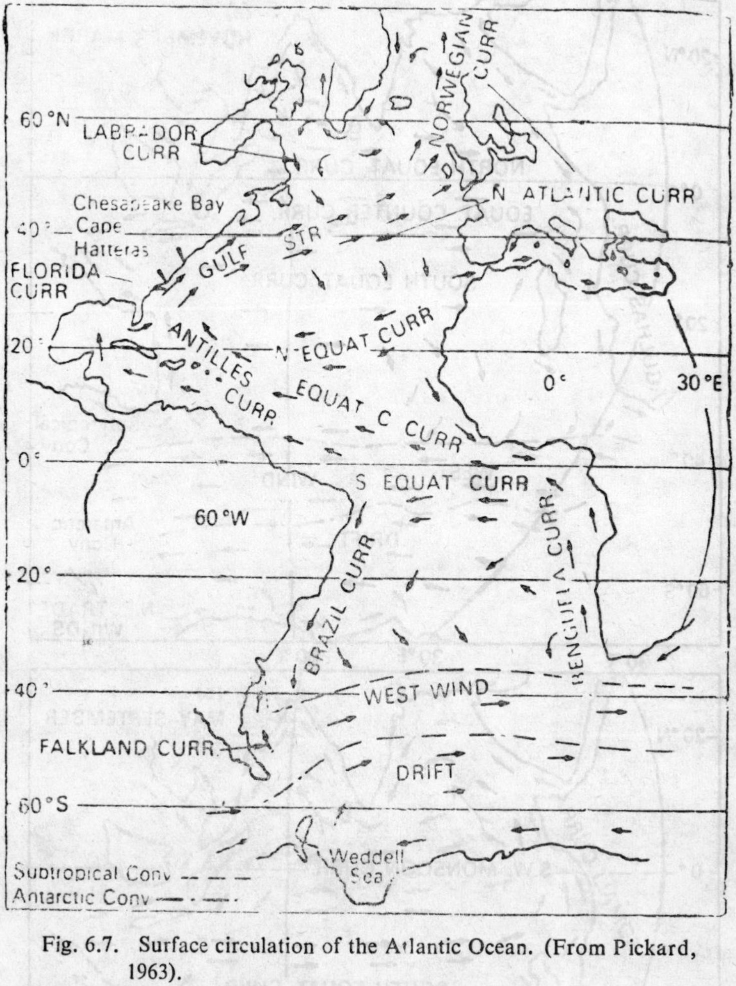

Fig. 6.7. Surface circulation of the Atlantic Ocean. (From Pickard, 1963).

to March, when northeast Trade Winds (northeast monsoon maintain the North Equatorial Current, which is close to the equator (with the change to the southwest monsoon from April to September, the North Equatorial Current reverses direction and the Equatorial Counter Current is no longer distinguishable).

The Trade Winds not only drive the equatorial currents but also their continuation as vast gyres that encircle the subtropical regions : hence the evocative phrase 'the backbone of

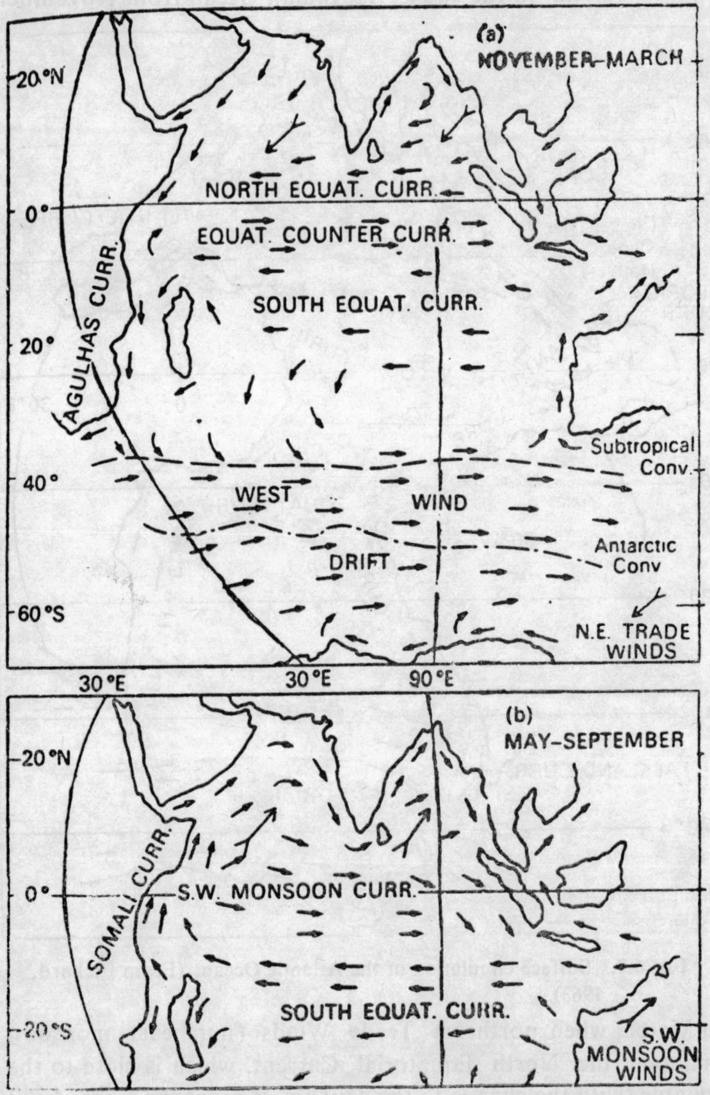

Fig. 6.8. Surface circulation of the Indian Ocean. (From Pickard 1963).

the circulation', seems justified. These oceanic gyres thus turn clockwise in the northern parts of the Atlantic and Pacific (the Asiatic land mass precludes the formation of a gyre in the northern Indian Ocean) and anticlockwise in the southern parts of all three oceans. The gyres overlie the Central Water Masses and, indeed, may extend into these waters.

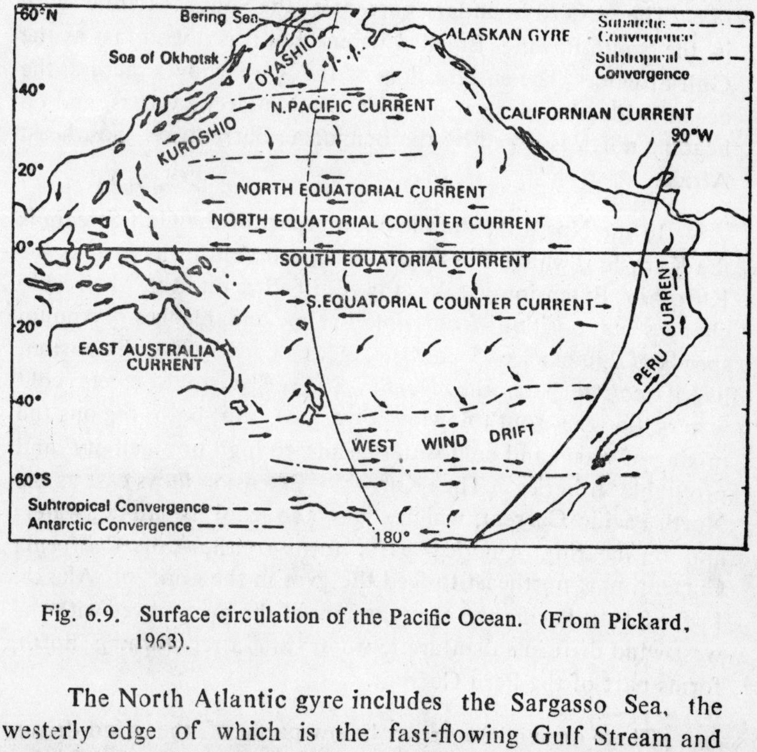

Fig. 6.9. Surface circulation of the Pacific Ocean. (From Pickard, 1963).

The North Atlantic gyre includes the Sargasso Sea, the westerly edge of which is the fast-flowing Gulf Stream and largely the continuation of the Florida Current that springs from the Gulf of Mexico. Speeds in the Stream, which reach 250 cm/sec (5 knots), are among the highest recorded in oceanic currents. Detailed surveys have shown that the Gulf Stream consists of narrow streams that trace a sinuous northeasterly path, and occasionally great eddies are detached to turn in the Sargasso Sea for months at a time. The Stream breaks away from the American coast at Cape Hatteras and near the Grand Banks of Newfoundland, where it broadens into the North

Atlantic current, it impinges on cold water from the Labrador Current. Part of the North Atlantic current moves eastward across the ocean and then turns south east Spain to complete the gyre in the North Equatorial Current : the other part flows northeasterly and eventually becomes part of the circulation of the Greenland, Norwegian and Arctic Seas.

The western boundary current of the South Atlantic gyre is the south-flowing Brazil Current but it is not so fast as the Gulf Stream. The eastern flow of the gyre moves across the ocean with the west wind drift in subantarctic waters, and on heading north is edged by the Benguela Current off Southeast Africa.

In the North Pacific the counterpart of the Gulf Stream is the Kuroshio, which flows past Japan and then turns east as the Kuroshio Extension. Like the Gulf Stream, the Kuroshio changes its position from time to time and reaches maximum speeds of 250 cm/sec (5 knots). Again, the Kuroshio Extension, like its counterpart, the North Atlantic current, meets cold waters (in the south flowing Oyashio). In both regions the mixing of warm and cold waters leads to high productivity and profitable fisheries. The Kuroshio Extension flows east as the North Pacific Current, which edges the cool subarctic waters and on nearing America parts turns south, as the California Current, part northeast to feed the gyre in the Gulf of Alaska. In the South Pacific the easterly flow of the gyre moves with the west wind drift in subantarctic waters and after turning north, forms part of the Peru Current.

There are undercurrents below parts of the wind-driven circulation. Below the Gulf Stream there is a counter-flowing undercurrent with speeds of 9-18 cm/sec but it is not always present. This was discovered by plotting the movements of Swallow neutral-buoyancy floats, which also revealed that deep currents in the North Atlant c are not invariably sluggish and steady. In the Bermuda area, where slow deep currents were expected, the flow at a depth of 2,000 metres averaged 6 cm/sec (maximum 20 cm/sec) and at 4,000 metres the average was 12 cm (maximum 42 cm/sec). Off Portugal measurements at

1,500 to 4,000 metres indicated currents of 1 to 2 cm/sec (maximum 5 cm/sec), which seems more in keeping with earlier estimations of deep water movement in the Atlantic.

Undercurrents are prominent below the equatorial systems and the first was discovered in the Pacific, where, 'The Equatorial Undercurrent is almost as large in volume transport as the North or South Equatorial Currents and it lies only 100 metres or less below the sea surface along the equator. Yet it was not discovered until 1952.' The outstanding discovery from current surveys, which was announced by Cromwell, Montgomery and Stroup, '...was that near the equator and embedded in the well-known westward flowing South Equatorial Current, there was a previously undiscovered eastward flowing current between 70 and 200 metres depth'. An intensive study by Knauss of Scripps Institution of Oceanography revealed many details of the current. It is a thin ribbon, only 0 2 km thick, but 300 km wide, extending from about 2°N to 2°S. This ribbon certainly stretches over the eastern half of the Pacific and may well be 12,000 km in length. The speed in the eye of the current is up to 150 cm/sec and the volume of water transported approaches that of the Gulf Stream. 'In other words, it is a major ocean current which was unknown before 1952 and had not even been predicted by theory' (Pickard, 1963). There are also Equatorial Undercurrents in the eastern half of the Atlantic Ocean and the Indian Ocean (from about 60°E to 90°E). The speed of the Indian Equatorial Undercurrent is only about half that of its counterpart in the other two oceans.

The wind-driven circulation has profound effects on the lives of deep-sea animals. The upward transport of nutrient salts from subsurface levels to the productive, plant-bearing surface layer is quickened along the edges of swift current (e.g. the Gulf Stream), or along lines of current divergence, or slowed in the great subtropical gyres that turn above a more or less permanent thermocline. As we shall see, the outstanding patterns of plant productivity are reflected down to all levels is the ocean, certainly affecting the abundance and the community structures of deep-sea faunas. Each night, the productive surface waters are the feeding grounds of diverse that move up from the upper mid-waters. These waters also form the nursery grounds for the young stages of many pelagic and some bottom dwelling deep-sea animals. During their planktonic existence the young may be widely dispersed by the lively wind-drived circulation and eventually some will grow into recruits for their parental populations. The quickest parts of the circulation may well carry the young away from the parental grounds but some, as they metamorphose and descend, may recurr in the undercurrents. Gyres and swirls will, of course, tend to keep planktonic forms within the confines of their circulation. If the ocean may be likened to a marine organism. the living units must above all evolve the right kind of organization and staying power to survive in the great organism's circulation.

Light in the ocean

Short-wave energy streaming from the sun heats and illuminates the ocean. Light penetrating the sea is both scattered and absorbed. Both processes are involved in the submarine attenuation of light which increases exponentially with depth and varies with the wave length. Virtually all of the energy in waves shorter than the visible (ultra-violet) or longer than the visible (infra-red) is absorbed in the top metre of the sea. At a depth of 10 metres most of the energy is in the blue-green part of the spectrum, though there is still a trace of visible red light. Thus, reddish organisms appear dark or black at depths below 10 metres, for at these levels there is little or no red light to be reflected by there organisms. It is thus not altogether surprising that diverse midwater invertebrates have adopted vanishing coats of red pigment.

Blue-green rays are still detectable near a level of 1,000 metres in the clearest oceanic waters, but no more than 5 per cent of the light energy penetrates to 50 metres. Even so, the light is strong enough for plant photosynthesis at depths down to 150 metres in the most transparent regions, where between

50 and 100 metres there is also enough light to allow divers to work.

The twilight zone of the ocean thus falls between the productive (euphotic) zone and the greatest reach of blue-green sunlight, which is probably close to 1,200 metres. The diversity of animal life in this upper midwater (mesopelagic) zone, greatly exceeds that in the vaster, deeper midwater (bathypelagic) zone, where there is no trace of sunlight. The twilight zone is full of animals that migrate daily to the more productive waters that move overhead. The most complex animals, notably decapod crustaceans, squid and fishes, have very sensitive eyes, able, more probably, to detect daily changes in light level and so regulate their vertical migrations. The twilight zone is also full of luminescent animals and with few exceptions, the light produced is blue-green and close in wavelength to the ambient solar light. Furthermore, eyes are most sensitive to blue-green rays, which seems almost inevitable in this marvellously integrated twilight world.

Just below the surface, where the sun's rays are reflected and refracted, the field of light is diffused and complex. At deeper levels most of the radiation is downward and approaches the zenith with increasing depth. At depths between 300 and 400 metres in clear oceanic waters the field of light is likely to be symmetrical around the zenith, and the polar diagram looks rather like the cross-section of a fish. Indeed, by matching this ambient field of light with the light of their own photophores, certain fishes of the twilight zone vanish from the view of predators.

Life zones and life forms

It must now be evident that water masses and pelagic forms of life are disposed in horizontal strata in the ocean. Pelagic organisms drift and move in submarine space : they are adapted for life at all levels between the air/sea interface and the interface between sea and substrate. Benthic organisms live on the ocean floor.

For many millions of years, certainly from Cambrain times (600 million years ago), the vast pelagic and benthic living spaces have been the scenes for a succession of .evolutionary plays known as adaptive tradiations. Today, according to Thorson (19/1), there are about 160,000 species of marine animals, of which 98 per cent are benthic and 2 per cent pelagic. Perhaps about 5 per cent of the total number of marine animals live in the deep sea, where the proportions of pelagic forms is perhaps greater. At all events this proportion certainly differs in large groups with both pelagic and benthic representatives. There are considerably more benthic, deep-sea cnidarians, polychaete worms, crustaceans and molluscs than of related pelagic forms. But pelagic deep-sea fishes are about as diverse as bottom-dwelling kinds.

Pelagic organisms, which range in size from bacteria to the largest whales, are broadly classed as planktonic or nektonic forms. Plankton (from the Greek *plankton*, neuter of *planktos*, wandering) consists of diverse organisms that are non-motile or have limited powers of movement : they thus drift, or tend to drift, with the sea. Pelagic bacteria are classed as bacterio-plankton, pelagic plants as phytoplankton and pelagic animals as zooplankton. Bacteria and the plants are microscopic forms, and so are some of the animals, such as the Protozoa and the early larval stages of invertebrates. Most adult members of the zooplankton range in size from millimetre-long copepods to large jellyfishes (e,g. *Cyanea* over a metre in diameter. The most ubiquitous forms are copepod crustaceans : other important groups are cnidarians, ostracod crustaceans, pteropod molluscs, arrow-worms, and the pelagic tunicates.

Members of the nekton (from the Greek *nekton*, neuter of *nektos*, swimming) have sufficient powers of locomotion to make their way against currents. Cetaceans, seals, turtles, fishes and squid are classic forms of nektonic animals. But where does the plankton end and the nekton begin ? The commonest deep-sea fishes belong to the genus *Cyclotbone* and the smallest species range from 25 mm to 50 mm in length. Several species of euphausiid shrimps fall within this size range, or even exceed it, and the group as a whole is often regarded as planktonic. Large euphausiids must have just as good powers of movement as small *Cyclothone*. Indeed, certain of the deep-sea prawns, which are larger than euphausiids, prey on *Cyclothone*. It seems reasonable then, to have a category called the micronekton (Marshall, 1954), which would consist largely of decapod crustaceans and the smallest cephalopods and fishes. At all events, nektonic animals are larger than say, 15 mm and they belong to groups with plenty of muscle. Such animals, which need to maintain large, energy-consuming muscles, are mostly omnivores or carnivores. Many planktonic animals, notably copepods and the pelagic tunicates, are entirely or largely herbivorous.

Animals that move or float on the sea surface have the collective name pleuston. The movers are water-striders (*Halobates* spp.) of the insect order Hemiptera : the floaters have some kind of gas-filled buoyancy chamber and include various cnidarians (e.g. *Physalia, Velella, Porpita* and certain sea-anemones), stalked barnacles, purple snails (*Ianthina* spp.) and nudibranch molluscs (e.g. *Glaucus*). Their headquarters are on the warmer parts of the ocean, which seem very remote from the levels of deep-sea life. Each day, though, many animals of the upper mid waters move into the surface layer and some, as we shall see, feed close to the surface, where they may well fall prey to the cnidarians, especially to *Physalia*, the Portuguese man-of-war.

Organisms that live in the uppermost few centimetres of the sea form the neuston. There are bacteria, tiny species of phytoplankton (nanoplankton) and diverse planktonic animals. Relatively few species of neustonic animals stay near the surface throughout the entire day, but from time to time they are joined by others, especially during the night. In subtropical and tropical regions many members of the neuston (e.g. pontellid copepods, arrow-worms, pelagic tunicates and larval stages of crustaceans and fishes) may be suffused with blue or purple pigments (David, 1965; Herring, 1967). Again, life close below the surface seems remote from that in the depths, but young stages of upper midwater (mesopelagic) animals, such as larval sergestid prawns and lantern-fishes, may form part of the neuston. Furthermore, neuston nets catch adult lantern-fishes (e.g. *Gonichthys coccoi*, *Symbolophorus veranyi*, *Myctophum punctatum*), which evidently take some neustonic food during their nightly stay in the surface waters, Indeed, land-dwelling insects that have been blown out to sea and eventually come to rest on the surface film may make up much of the food of certain lantern-fishes. Concerning the neuston, one must remember though, that it is concentrated in a very thin layer and must thus be no more than a subsidiary source of surface-dwelling food for nektonic animals, particularly in tropical and subtropical waters. In these regions the overall impression one gains of the near surface (0-10 cm) zone (compared to the layer from the surface to 30 metres) is that of a poor feeding ground with a rather low density of invertebrates and fishes.

The uppermost waters of the open ocean are known as the epipelagic zone. Here, in the subtropical and tropical belts is the swim of flying squid, flying-fishes, skippers, sauries, grey sharks, thresher-sharks, mako sharks, tuna-fishes and billfishes. The sharks, tunas and billfishes are Gullivers compared to the Lilliputian fihes in the waters of the deepsea world. Many thousands of the largest lantern-fishes and stomiatoid fishes are needed to equal the weight of a large bluefin tuna or a blue marlin.

What is the lower limit of the epipelagic zone? Tuna fishes live in the upper 200 metres and are sometimes concentrated near the 'seasonal' thermocline but this varies in level between about 10 to 250 metres. Planktonic plants live in the surface layer and sunlight is strong enough for photosynthesis down to 150 metres in the clearest oceanic waters. Not far below this depth small luminous fishes with silvery sides (e.g. the stomiatoids such as *Bonapartia*, *Vinciguerria* and *Valenciennellus* begin to appear in nets towed at successively deeper levels below the surface. We shall accordingly, and somewhat arbitrarily, take the lower limit of the epipelagic zone to '' about 150 metres, which is also close to the mean depth of the outer edge of the continental shelf where shallow seas give way to deep seas.

If the mesopelagic zone is regarded as a twilight environment, then the transition between this zone and the dark underlying bathypelagic zone is the threshold of light. Measurements of underwater light by Clarke (1970) and his colleagues have shown that in very clear oceanic waters sensitive instruments will detect sunlight down to about 1,200 metres. There are not enough measurements to give a reliable estimate of the mean depth of light penetration in the entire oceanic province, but this is likely to be close to 1,000 metres. In subtropical and tropical waters the mesopelagic zone contains the main thermocline, but everywhere below this zone temperatures at a given level are virtually uniform and range between 5°C and negative values. The bathypelagic zone is thus uniformly cold and dark (apart from flashes of bioluminescence). Considering that the headquarters of deep-sea animals are below the warm oceanic belts, what are the biological features of the twilight-thermoline-mesopelagic and the dark-cold-bathypelagic zones. Full treatment of the mesopelagic and bathypelagic faunas comes later in this book : here we look briefly at copepods and fisnes, the two ends of oceanic food chains. At depths below 1,000 metres there is a distinct and rather diverse bathypelagic fauna of calanoid copepods, distinguishable from mesopelagic and epipelagic kinds. some of which are related forms (Grice and Hulsemann. 1965). Concerning the fishes, the transition between the mesopelagic and bathypelagic zones is marked by certain dark-coloured species of lantern-fishes of the genera *Lampadena* and *Taaningichthys*. Lantern-fishes, most of which migrate daily between their daytime levels and the upper waters, are classic mesopelagic animals. Ceratioid angler-fishes, dark-coloured species of bristle-mouth fishes (*Cyclothone*), whale-fishes and gulper-eels, which are non-migrators, are typical dwellers at bathypelagic levels. This is not to say that they are entirely confined to the bathypelagic zone. The vertical ranges of certain black species of *Cyclothone* and angler-fishes (e.g. *Melanocetus* and *Cryptopsarus*) may extend into lower mesopelagic levels. The transition between the two midwater zones, whether physical or biological, is by no means abrupt.

Besides housing many kinds of vertical migrators and the reflections of some as sound-scattring layers (which do not occur at bathypelagic levels), the mesopelagic zone is also, as already stated, full of luminous animals. At depths below 1,000 metres luminous flashes become much less frequent (Clark, 1970). Indeed the specific diversity of life and the amount of life (biomass) are much reduced at bathypelagic compared to mesopelagic levels. For instance, for every species of the bathypelagic fish fauna (about 150 spp.) there are six mesopelagic species. The biomass of zooplankton decreases markedly between the surface layer and 1,000 metres but at deeper levels the decline is less rapid. Beyond 2,000 metres there is a gradual and steady exponential fall in bimass (Vinogradov, 1968).

In the deeper parts of the bathypelagic zone below 4,000 metres there seems to be a distinct abysopelagic fauna, which contains, inter alia, members of the copepod family Spinocalanidae. Species of this family live also near the deep-sea floor as part of a benthopelagic fauna, which ranges from these millimetre-sized copepods to certain fishes (macrourids) and cirrate octopods that reach lengths of about a metre. Part-time members of the benthopelagic fauna include pelagic forms (e.g. medusae and jellyfishes) and also some benthic species (e.g. amphipods).

Animals of the deep-sea floor, like their benthic relatives on the shelf, live partly or entirely above the interface between sea and land (epifauna) or within the upper layer of sediment (infauna). The infauna consists largely of minute (melobenthic) forms (e.g. forminifera and nematode worms) and small species of burrowing bivalves, crustaceans and polychaete worms. The epifauna is not only represented by members of the taxa that flourish on the continental shelves, but there are others that are predominantly deep-sea forms, notably the glass-sponges, beard-worms (Pogonophora) and elasipod sea-cucumbers.

Down the continental slope and rise and moving deeper into the ocean basins there is a decline in the biomass and species diversity of the benthos. The decline is most marked in most of the epifauna but is much more gradual in the relatively small forms of the infauna. Some specialists differentiate a bathyal fauna, extending down the slope to about 2,000 metres, from an abyssal fauna at deeper levels. Critical survey of these faunas is beyond the scope of this book.

7

Island Environments

Introduction

The study of geography of island biota, or island biogeography, has strongly influenced understanding about patterns of life on the Earth. Major advances in thought about the processes of evolution resulted directly from observations of island organisms. Darwin's work on the Galapagos Islands, Wallace's work on the Malay Archipelago, and Hooker's exploration and observation of South Pacific islands all inspired significant contributions to the development of understanding of various physical and biological features of the Earth.

This chapter explores the contributions that the study of islands and their biota have made to biogeographical thought. In the first part of the chapter, the most commonly recognized island, that is, isolated land features in the midst of the sea, will be considered. These types of islands have a combination of physical and geographic features that make them natural laboratories. Secondly, some of the principal biotic features of islands will be examined. Modern theories of island biography have stimulated research, and some of these ideas will be explored. Finally, other forms of islands such as mountain tops, deserts, and refuges of species will be described. Considering their small size, islands play an important biotic role on the Earth.

The physical geographic setting of islands

By definition, islands are small bodies of land surrounded entirely by water. Bodies of land as large as or larger than Kalaatdlit Nunaat (Greenland) are generally thought of as New Guinea, Borneo, Malagasy (Madagascar), Baffin, Sumatra, Honshu, and Great Britain. The land area of the Earth's islands is about 4-5 × 10 km² or about 3% of the land area of the Earth. Thus, as in the case of lakes and rivers, islands make up a relatively small area compared with their biogeographic importance to the Earth.

The processes of sea-floor spreading and plate tectonics were described. The geophysical processes that are involved in the movements of the continents and the creation of the sea floor also explain the creation of the two major types of islands, continental islands and ocean island. Continental islands have a geologic structure that is similar to that of continents and their origin can be traced back to a past connection with a continental land mass. They contain granite as well as rocks of sedimentary, igneous, or metamorphic origin similar to those of continents. These may be recognizable in the surface rock structure as in the British Isles, or they may be obscured by lava and volcanic ash as in the island are which occurs on the margin of the continental plate in the destructive zone. The latter includes the Japanese islands and the Malay Archipelago.

Oceanic islands do not have granite, sedimentary rocks, metamorphic rocks related to continental geologic structures. They are volcanic in origin and made up of basalt. They form under three circumstances : along oceanic ridges where large amounts of basalt are moving to the surface of the ocean crust; at hotspots; and along subduction zones within island arcs. Examples of these island include Surtsey, Tristan da Cunha and other islands of the Mid-Atlantic Ridge, and the Galapagos Islands (although, as discussed later, there is dispute about this fact) on the margin of the Cocos and Nazca plates; the Hawaiian and Emperor seamounts in the Pacific; and many volcanic islands on the ocean side of the destructive zone as in

Fig. 7.1. A map of many of the islands of the Earth.

the Kurile are and the western Aleutian arc. Note that from island arcs there may arise islands that are either continental and complicated by volcanism or are in a relatively simple form as oceanic volcanic islands.

From a biogeographic point of view, oceanic islands can be subdivided into volcanic high islands, atolls, and raised limestone (Scott & Rotondo 1983). Atolls consist entirely of coral. Coral atolls are especially common in the Pacific where there are 309 known to exist. The Indian Ocean has 73, the Caribbean 26, and the Atlantic jast 2, Rocas near the coast of Brazil and Bermuda (Thomas 1963). Islands can of course be extremely complicated mixtures of granitic, basaltic, and carbonate materials (Uyeda 1978).

Also relevant to an understanding of island life are the facts surrounding the climatic perturbations that have been associated with the advances and retreat of continental ice sheets. Many possibilities exist. Some present-day islands, such as the British Isles, were formerly connected via dry land to the continent during the peak of the ice age, and became islands only when the ice melted and sea level rose. Other islands, such as the highest seamounts of the Pacific, were probably islands the peak of glaciation and became inundated at the end of the glacial period. Islands located in higher latitudes were subject to extremely cold climatic conditions during the last glacial episode, and those not influenced by warm currents were probably stripped of most of their flora and fauna. Thus, long-term climate variations have had large influences on islands both because of the effect on sea level and the direct effect on the climatic conditions of the island. The problems associated with climatic change are magnified on islands because they generally exhibit little changes, organisms cannot readily move to a more favorable location that has a suitable range of climatic variables. Organisms confronted by new circumstances and unable to escape must either adjust or suffer decreases and the likelihood of extinction.

Climate and sea level changes that accompanied the glacial ages are not the only variables that affect island biota. **Hurricanes and their accompanying storm surge can inundate**

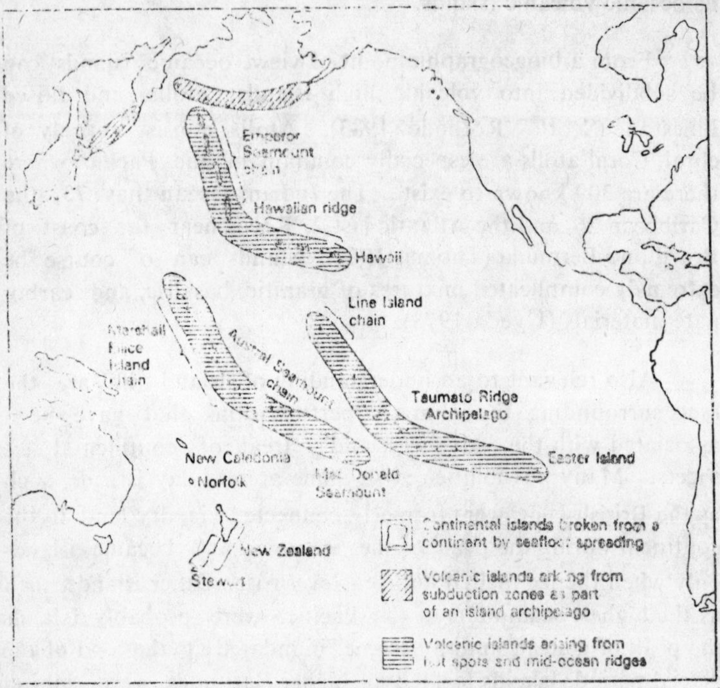

Fig. 7.2. Map of the Pacific Ocean depicting the three major types of physical origins of islands.

low islands and atolls. The frequency and magnitude of such events is a matter of probability that is in part influenced by geographic conditions. Regional climatic variations are moderated by the mass of water which surrounds islands, but even this generalization has to be critically evaluated in the light of the evidence. Some islands in tropical waters are subject to remarkable climatic variation. The El Nino event of 1982-83 caused immense disruption to marine and terrestrial habitats, both proximal and distal to the Tropics where the unusually high water temperatures were recorded. Central Pacifie seabird abandoned atolls such as Christmas Island (Schreiber &

Schreiber 1984). Breeding grounds were submerged by high tides and runoff from extremely heavy rains induced by El Nino. Adult birds apparently had difficulty fishing in rain and temporarily migrated away from their usual habitat. They

Table 7.1

Area and relief of New Zealand and neighboring islands

Islands	*Area (km^2)*	*General relief*
South Island	148,700	50 peaks over 2750 m
North Island	113,400	3 peaks over 2000 m
Stewart	1,720	Granite, 3 peaks, up to 987 m
Chatham	950	Cliffs to 286 m
Auckland	600	Volcanic, one peak 615 m
Macquarie	118	Volcanic. 436 m glacial lakes
Campbell	113	574 m glaciated
Antipodes	60	Volcanic, peak of 406 m
Kermadec	30	Volcanic, peak of 542 m
Three Kings	8	Volcanic
Snares	2.6	Granite, cliffs to 197 m

returned when the El Nino event ceased. Thus, island life is vulnerable to short-term weather events as well as long-term climatic changes.

The range of climatic conditions of an island is determined by the elevation of the highest mountain peaks. Island area is normally a good indicator of elevation and hence climally a good indicator of elevation and hence climatic heterogenity. However. closer examination of one group of islands suggests complexities in physical variables found among islands. New Zealand and neighboring islands comprise one group of islands that is reasonably well explored and catalogued.

Generalizations are very difficult to make from this table. The largest islands have the highest peaks, though smaller islands have no consistent pattern. For example, Campbell Island, the fourth largest island of the group has lower peaks than Kermadec Island of the ninth largest of the group. Chatham is over 30 times larger than Kermadec but has peaks half as high. Note also that the islands vary significantly in geologic structure with the largest and smallest being continental or granitic and most of the other islands being volcanic in origin.

This group of islands has a wide range of temperatures. It is important to note the correspondence between increasing latitude and decreasing temperatures for the group. Being located in the Southern Hemisphere, these islands exhibit a more consistent change of temperature with latitude than islands located in the Northern Hemisphere. This is mainly a function of the fact that the Southern Hemishere is dominated by water and contains only one-third of the Earth's land mass.

If any generalizations can be made about islands, they would be along the following lines of reasoning. Regarding geologic conditions, most islands are the product of some of the Earth's most violent forces. Many are born of fire and a shower of ash and rock they often exist within the context of enormous volcanic eruptions and explosions. Perhaps as devastating for land life is the fact that oceanic islands gradually subside below sea level. The forces that create islands also pose secondary threats to islands. In the ocean, earthquakes associated with plate tectonics set up powerful and exceedingly fastmoving tsunami waves. The amplitude of tsunamis becomes enhanced on slopes and produces a high wave of water that can sweep over some low-lying islands. The effect of tsunamis depends upon the angle of the offshore slope. Atolls, for example, have such a steep slope that despite their low elevation, the tsunami wave has no opportunity to build up. Tsunamis cause their most sever damage along coasts of gently sloping continental shelves and within embayments (perhaps on islands). Regarding climatic conditions, in tropical and subtropical seas, islands are subject to devastation by powerful tropical storms (hurricanes, typhoons, or willy-willies). These storms in combination with a storm surge can also sweep over a low-lying island. In sum, islands are proportionately more vulnerable to physical and climatic disruptions than larger land masses.

Table 7.2

Climatic data for New Zealand Neighboring Islands

Island	*Latitude S*	*Mean temperature (°C)*			*Mean daily sun(%)*	*Annual precipitation (mm)*
		Annual	*Jan.*	*July*		
Kermadec[a]	29° 15′	19	21	16	47	1504
North Island						
Kaitaia[a]	35° 04′	15	19	11	47	1370
Wellington[a]	41° 17′	12	15	8	45	1224
Chatham[a]	43° 57′	11	14	8	33	838
South Island						
Dunedin[a]	45° 52′	11	14	6	40	940
Invercargill[a]	46° 25′	10	13	5	37	1063
Stewart[b]	47° 20′	n/a	13	4	n/a	n/a
Auckland[b]	50° 50′	8	11	6	n/a	1450
Campbell[b]	52° 33′	7	9	5	15	1441
Macquarie[a]	54° 30′	5	7	3	n/a	952

Sources : [a] Rudloff 1981, [b] U S Dept. of Commerces 1968.

Biotic features of islands

Given the above physical features of island, is easy to understand why islands have been simply subdivided on the basis of habitat and biota into low islands, such as atolls and sand cays, and high, rugged islands (Carlquaist 1967). Low islands have little more than coastal habitats. Their biota is drawn from an inventory of widespread coastal species, but the particular set of species on different islands, even in close proximity, may vary significantly. For example, atolls of the northern Line Islands have similarity indices of indigenous plant species ranging from 36.4 to to 78.2 despite their close proximity (Wester 1985). In contrast, high islands, in addition to having various elements of the more or less cosmopolitan coastal biota, also have complements of inland biotas that are strongly influenced by each island's geologic history, physical geographic relationships, location, and human history. The features of high islands have stirred the imaginations of naturalists such as Darwin. Wallace, and Hooker, as they studied island life.

On the voyage of the *Beagle*, Darwin visited a large number of islands that had a variety of climates and environmental conditions. A broad outline of the distribution of life forms already existed prior to Darwin's time. Under similar climatic and edaphic conditions it was expected that the same forms of species would occur. Instead, Darwin encountered strong similarities between the Galapagos and South American biota living in entirely different environments and large differences between species living in the Galapagos Islands and other places exhibiting similar environments, such as the Cape Verde Islands. In the Galapagos Islands Darwin had been collecting and mixing together specimens of tortoises from each island. He then learned from a local resident that tortoises from each island were clearly recognizable from each other. Darwin wrote that he never imagined that islands so close and similar to one another would be inhabited by such distinct fauna. Fortunately, some of the bird specimens that Darwin collected were catalogued by island. The Galapagos finches formed the corner stone for the idea of natural selection (Lack 1947).

Alfred Russel Wallace (1892) wrote the book *Island Life* to summarized and popularized his two-volume work *Geographical Distribution of Animals*. Wallace mode some extremely important observations about the distribution of organisms on the Earth. For example, he recognized the biotic similarities between Japan and England that were 21,000 km apart and the differences between Australia and New Zealand that the separated by a distance of 2,100 km as well as the even more striking case of biotic dissimilarities on the islands of Bali and Lombok in the Malay Archipelago that are separated by a distance of only 23 km.

Wallace also pointed out the case of the interesting group of organisms in eastern North America composed of pines, oaks, sumachs, and magnolias. However, traveling a distance of only 80 km from Florida to the Bahamas, one encounters an entirely different vegetation that is akin to the Caribbean and tropical floras. Of this situation, Wallace stated, "Between frigid Canada and sub-tropical Florida there are less marked differences in the animal productions than between Florida and Cuba or Yucatan, so much more alike in climate and so much nearer together". These are many remarkable examples of similarities and differences between various combinations of continents and islands led Wallace to conclude that "mere distance is one of the least important of the causes which have determined the likeness in the animals [and plants] of different countries".

Wallace stressed the interdependence of organic and inorganic nature. To Wallace, "the long series of past geological change", "the periodic variation of terrestrial climates", "the fortuitous action of storms and currents for conveyance" and "the endlessly varied actions and reactions of organised beings on each other" were all topics to be fully explored and reconciled in the study of the distribution of living organisms, As will be shown, modern theory does not place the study of the island biogeography within as broad a context on Wallace considered over a century ago. "Islands possess many advantages for the study of the laws and phenomena of distribution. As compared with continents they have a restricted area and definite boundaries, and in most cases their geographical and biological limits coincide." These words place islands into a framework that modern theorists have attempted to use to develop abstractions and models to test island evolutionary concepts.

The equilibrium model

The most influential contemporary work on the topic of island biogeography is the book entitled *The Theory of Island Biogeography.* Following is a discussion of the hypothesis of the equilibrium model of continuous immigration and extinction as presented by MacArthur and Wilson. The predicted outcome will then be tested using empirical data of New Zealand and neighboring islands. The theory considers two variables, immigration and extinction, and their theoretical interaction.

The immigration of species per unit of time is hypothesized to proceed at a decreasing rate because as species become established on an open habitat, fewer immigrants will belong to new species. Species that are rapidly dispersed become established first, causing a rapid decrease in the immigration rate while the more slowly dispersing species cause a continuously diminishing rate of immigration. The described curve is a concave immigration curve.

Extinction has been hypothesized to proceed at an increasing rate because the larger the number of species the more likely it is for one of the species to become extinct due to decreasing population size. The extinction curve is exponential : "reasoning that the combination of diminishing population size and increasing probability of interference among species will have an accelerating detrimental effect". Together, the immigration and extinction curves form the basis for the equilibrium model.

Several predictions arise from the equilibrium model when it is applied to various sizes of islands at various distances from each other. If the rate of immigration is great or the rate of extinction small, it is hypothesized that the number of species will be large. Two islands will contain a comparable number species either by being near to the source region (and immigration is higher) or by being larger (and the extinction rate is lower). Conversely, it is hypothesized that the number of species will be lower either if the island is distant from a source region and the immigration rate is small, or an island is small and the extinction rate great. The discription of the near-far and small-large island hypothesis is illustrated by the curves. MacArthur & Wilson predict that "the logarithms of the

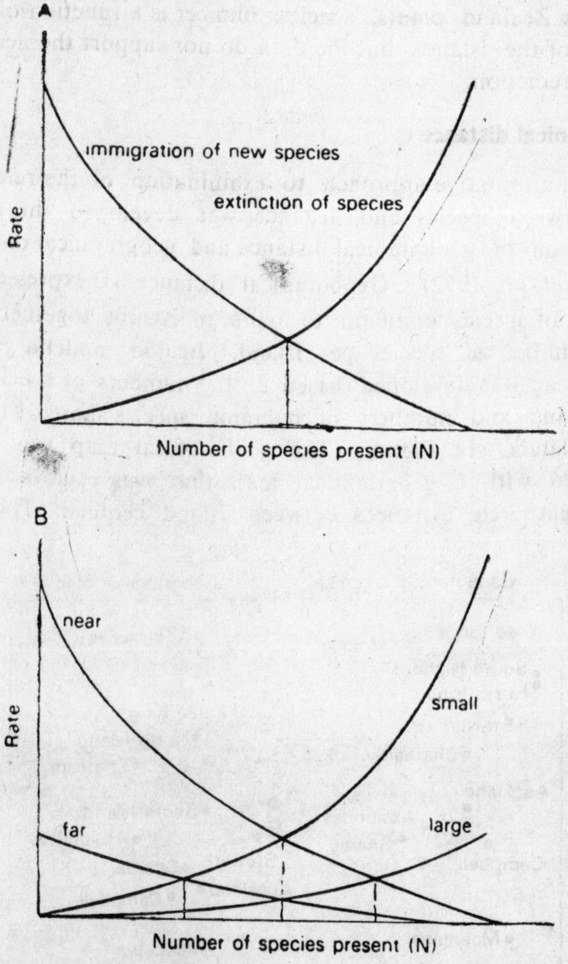

Fig. 7.3. Two models by MacArthur and Wilson that have simulated thought about island organisms.

species number increase with area more rapidly on distant islands than one near islands." One test of the prediction was accomplished by using the relatively well-described flora of New Zealand and neighboring islands (Tobler *et al.* 1970). Many problems arose in testing the prediction. For example, what is the distance that separates "near" and "far" islands ? For New Zealand plants, species, number is a function of area of each of the islands but the data do not support the near-far island prediction.

Geobotanical distance

An alternative approach to examination of the relationship between species and distance was developed through a comparison of geobotanical distance and geographical distance (Tobler *et al.* 1972). Geobotanical distance is expressed as numbers of species common to pairs of islands together with total number of species per island. In the model a geobotanical map was developed based on the numbers of species on each island and numbers of common species shared with all other islands, see Figure. The geobotanical map was then compared with a geographical map that was established by using great circle distances between island centers. The two

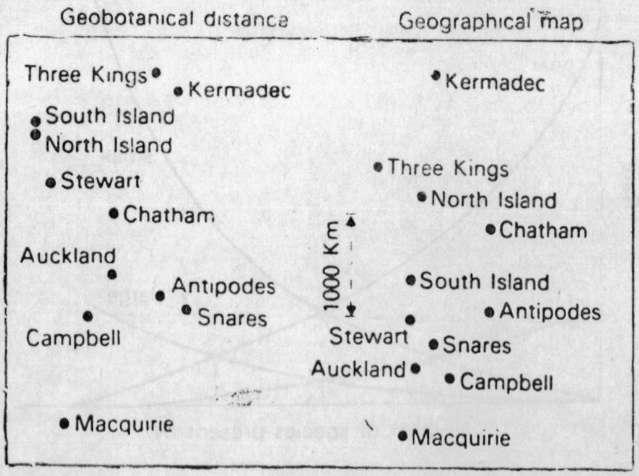

Fig. 7.4.

maps had about at 34% correlation between them. Severai other attempts at predicting similarities of dissimilaritics between islands were tried with more or less similar results (Tobler *et al.* 1970). The geobotanical distance model has problems that are similar to the equilibrium model. The problems fall into three major categories relating to the botantical data and environmental and biological assumptions that are held to be true for the models.

Problems of the Models

Botanical Data

Any attempt at empirical testing of island models relies on the quality of the knowledge of the floras and faunas of the islands and the successful application of the art of classifying organisms into categories. The taxonic data for New Zealand and neighboring islands were mainly drawn from Allan's *Flora of New Zealand* (1961). The obvious advantage of using data from a single flora by one anthor is the increased likelihood of uniform taxonomic treatment and steadfast attention to the distribution of plant species on various islands. The actual quality of taxonomic work or attention to the detalis of species distribution is not known. The data include only dicotyledonous species which comprise a minority of the plants, ranging from 33% of the species on Antipodes Island to 49% of the species on Stewart Island. The islands also have ferns, gymnosperms, and monocotpledons that were not included in Allan's *Flora*, Allan included only species that were regarded as indigenous. Unfortunately, the present knowledge of the flora and fauna of islands is not only imperfect but the composition of the indigenous flora cannot be known because of human modification (both biotic and physical) of the environment. No place, island or continent, has been spared from the changes that accompanied the spread of human culture and technology on the Earth.

Environmental Assumptions

The geobotanical and equilibrium models have assumed that the environment is static in both time and space. Yet, as the island are born of and under the influence of some of the most dynamic geologic and climatic forces found on the Earth. The islands may undergo considerable fluctuation is size as a result of sea-level changes, as well as eruptions and erosion on the islands. Wind and New Zealand case. Both currents are in a westerly or west to east direction within the region. Some large genera of Australian plants e.g., *Personia*, *Phebalium*, *Leptospermum*. *Haloragis*, *Epacris*) have representatives in the New Zealand region. Also, large New Zealand genera (e.g., *Hebe*, *Coprosma*) have several species in South America. In contrast, examples of plants suspected of east to west movement are uncommon.

Since the upper Cretaceous. changes have taken place that affects plants and their ability to migrate (Le Pichon 1961). For example, changes of distances between New Zealand, Australia, and Antarctica as well as concurrent changes in climate resulting from shifts in latitudes of each of the land masses have contributed to the present pattern of flora. Antarctica once supported tree species of *Nothofagus*, the southern bench, during the upper Cretaceous and Tertiary (Cranwell 1963). While the southern beech no longer lives in Antarctica, it is abundant in New Zealand. New Zealand's Antarctic affiliations illustrate former connections during apast geologic era rather than those of the present. New Zealand, through relicts such as *Nothofagus*, provides some glimpse of floras that are now extinct in their former habitats. The models do not include these aspects of geology and climate in their formulation.

Biological Assumptions

The biological assumptions of the equilibrium and geobotanical distance models are as static as the environmental assumptions. For example, the models assume that all species disperse with equal ease while in fact, there are large differences in dispersal abilities between species.

More significantly, the models fail to account for the most intriguing biological characteristic of islands, the unique species that are distinctive to them. From an initial immigrant (founder), differentiation and the development of an assortment of specialized characteristics evolve that allow the later generations to expand into unoccupied niches and habitats. The Galapagos finches are often cited as an example of adaptive radiation, though many questions remain unanswered about the Galapagos Islands. For example, recent geologic investigation suggests an above sea level (or subaerial) origin for the Galapagos rather than a below sea level (or submarine) origin, and an age of at least 3 Ma instead of no more than 1.5 Ma as formerly assumed (Hall 1983). The finding brings into question the possible means by which organisme have dispersed to the islands and the rate of speciation that has taken place, A far more extensive example of adaptive radiation has been that of the insects of Hawaii (Zimmerman 1970).

The Hawaiian Drosophilidae are particularly important because of the fact that genetic analysis was possible by studying giant chromosomes found in their salivary glands. Worldwide, there are about 1,500 species of *Drosophila*. Of the total, about a third of the species are endemic to the Hawaiian Islands. In addition, the genus *Scaptomyza*, another member of Drosophilidae, has about 300 species endemic to these islands. Thus, about 800 species of drosophilids inhabit Hawaii. One group, the picture-wing drosophila, has been studied intensely. It contains over 100 species, and phylogenetic relationships between the species of the Hawaiian Islands have provided insights into the various inter-island movements and the speciation that took place on single islands (Williamson 1981, Carson & Yoon 1982). Recent work on *Drosophila* on the island of Hawai has led to the development of the Kaneshiro model which accounts for speciation by considering both the founder effect and the behavioral differences that influence evolution (Giddings & Templeton 1983).

Other extraordinary examples of adaptive radiation include the birds of Hawaii which presently include 47 species and 38 subspecies belonging to 11 avian families (Shallenberger 1981), and lizards in the genus *Anolis* in the West Indies (Williams 1969, 1972). In summary, the discussion comes full circle and returns to Wallace's point of view. Understanding the distribution of the Earth's biota is not only a matter of considering distance and area, but rather requires examination of a whole host of factors. both inorganic and organic, that influence the movement and evolution of organisms.

Characteristics of the patterns of Island life

There are several distinct features which illustrate the general relationship between island biota and geologic origin, the dispersal capacities of island biota, and the phenomenon of radiation of species.

The relationship between biotic characteristics and physical origin of islands

Darwin recognized two physical types of islands, those of continental origin and those of ocean origin. He also recognized some important biotic distinctions between them. Continental islands are not usually remote from a continent and almost always contain some land mammals and amphibia. Wallace added the observation that continental islands may be subdivided into two groups, those of ancient origin and those of recent origin. Ancient continental islands are usually separated from the adjacent continent by a body of ocean water that is extremely deep. Their geologic structure does not differ from that of other continental islands. They often possess an abundance of both mammals and amphibia "but these are highly peculiar, almost all being distinct species, and forming distinct and peculiar genera or families." Ancient continental islands "are also well characterized by the fragementary nature of their fauna, many of the most characteristic continental orders of families being quite unrepresented, while some of their animals are allied, not to such forms as inhibit the adjacent continent, but to other found only in remote parts of the world."

By comparison, recent continental islands exhibit relatively simple features. They are connected to their adjacent continent by relatively shallow waters. Their geological structure is similar to that of the adjacent continent, as is their plant and animal life. In fact, their organisms and those of the adjacent continent are either identical or so closely related as to be clearly allied.

Dispersal

Island of oceanic origin have biota that is far more selected. To understand the patterns of oceanic islands life requires consideration of dispersal abilities of various.

The dispersal abilities of various organisms determines to a large degree which plants and animals are likely to end up on islands of varying isolation. Filtering, impoverishment, and disharmony are concepts that have been used to describe the variations in the kinds of plants and animals that are found on various islands. These concepts are similar in that they describe various aspects of the products of dispersal.

Filtering, impoverishment, and disharmony

Filtering is the reduction of the number of species, genera, or higher categories of plants and animals which takes place during dispersal. To animals which have poor dispersal abilities, such as terrestrial mammals, a body of water such as between Bali and Lombok is an impenerable barrier. For other animals such as birds and reptiles, the series of islands that make up the archipelago between the Orient-Sumatra and New Guinea-Australia act as a filter. In this case, since there are unique groups on either side of the archipelago the series of islands represents a two-way filter. Filtering leads to a reduction in the number of taxa, that is, impoverishment of island biota compared to the biota of the source region.

Impoverishment of oceanic islands floras was one of the characteristics first described by J.D. Hooker (Turrill 1964). In addition to the fact that oceanic islands generally have a smaller number of species compared with the same size of area of a continent, they also tend to have fewer species per genus or genera per family than similar areas on continents. Impoverishment can be illustrated by the flowering plants of New Zealand and neighboring islands.

Some authors refer to impoverishment as a form of filtering. The two phenomena are closely related because they are an outcome of dispersal. Impoverishment has a subtle twist to it. As in the case of New Zealand and neighboring

Table 7.3

Relative Dispersal Capabilities of Major Groups of Organisms

Plants

Terrestrial Algae and Fungi, because of small size good dispersers.

Ferns, mosses, club-mosses, horsetail psilotum; wind dispersed Primarily a function of size and density of spores.

Angiosperms

Large variability within families, but specific examples illustrate dispersal ability.

Unlimited, sea-water distributed plants (beach plants), some legumes, morning-glories, Scaevola, Cocos, and mangrove.

Excellent : seeds of fruits with special adhesive qualities (Pisonia, Bidens, Acaena, Boerhaavia), parachute-like mechanisms (asters. senecios, lettuces), tiny windborne seeds (orchids).

Very good, aquatic plants with seeds which can become imbedded in mud and carried on birds' feet (sedges, rushes).

Good, fleshy fruits with small seed eaten by birds (Vaccinium, Fuchsia, Hawaiian lobeliads).

Fair, fleshy fruits with medium-sized seeds that can be carried by birds (Myrsine, Araliaceae, Rubiaceae).

Poor, Large-seeded trees (oaks, plums, walnuts, mangoes, etc).

Conifers

Junipers, around 1,500 km from North America to Bermuda or Europe to Azores.

Podocarps, about 800 km from Fiji to Samoa or Tonga and may be more for New Zealand podocarps.

Pines. redwoods, araucarias, very short distances although pines crossed to Guadalupe Island from Mexico, at least 300 km.

Animals

Birds :more than 3,200 km (North America to Hawaii, South America to Tristan da Cunha).

Bats : more than 3,200 km (North America to Hawaii).

Land molluscs : more than 3,200 km (Polynesia to Juan Femandez Islands).

Insects and spiders : more than 3,200 km.

Freshwater fish : salt water gaps of only a few kilometers may be bridged.

Reptiles

Lizards, long distances for geckos.

Snakes, 800 km and more to Galapagos Islands.

Land tortoises, 800 km and more to Galapagos Islands.

Freshwater turtles, over 300 km to Madagascar.

Amphibians, probably only small distances.

Terrestrial mammals

Rodents, 800 km as in the Galapagos Islands.

Small mammals, 300 km in the case of Madagascar civets and insectivores.

Large mammals, perhaps 40 km with exception of semi-aquatic hippopotamus of Madagascar.

Source : Modified from Carlquist 1965.

Table 7.4

Impoverishment as measured by the ratio of species per genus for New Zealand and neighboring islands

Island	*Species per genus*	*Size (km²)*
South Island	4.55	148,700
North Island	3.14	113,400
Stewart	2.13	1,720
Chatham	1.49	950
Auckland	1.37	600
Macquarie	1.23	118
Campbell	1.53	113
Antipodes	1.25	60
Norfolk	1.18	34
Kermadec	1.35	30
Lord Howe	1 17	14
Three Kings	1.41	8
Snares	1.00	3

Source : Compiled from Allan 1961.

islands, the number of species represented within a single genus tends to be reduced on smaller islands, but the number of genera represented on these same islands is greater than would be anticipated given the stock of species within the source region. Thus, while there is a reduction of the absolute diversity of species, there is an increase in the relative diversity of taxonomic groups. One explanation for this phenomenon is that species within the same genus tend to exclude each other as they arrive on isolated islands. Perhaps, as described below, relatively few species of a genus retain their ability to disperse widely, while most variants of the same genus lose dispersal ability. If this were the case, then only a few representatives would have a reasonable chance of dispersing to isolated islands.

Filtering and impoverishment lead to enormous variation in the groups of species represented on clusters of islands with different distances from a major continent. The Channel Islands tend to have 2½ times more vascular plant species and 4 times more species of breeding populations of birds than equivalent-sized islands in the Azores. However both, groups of islands have about the same number of ferns (Williamson 1981). This example supports the idea that ferns are excellent dispersers and that vascular plants and birds are less able in their dispersal capabilities.

Another way of describing dispersal is in terms of inabitities, or disharmony, that result when certain groups that are usually together on continents become separated when they disperse to islands. Disharmony comes about when some groups of organisms fail to migrate to islands. Again, the above list of the dispersal abilities of various groups of plants and animals provides insight into the existence of disharmony of islands, especially distant and isolated islands. The species which arrive on such islands do so as a result of chance that is influenced by physical geography and biotic processes, such as bird migration and ability to travel long distances without loss of viability. The organisms that succeed at arriving on a barren island have a large number of barriers to overcome before they become established. However, once an organism has become established on an island, a number of general changes occur. These changes have made islands a source of fascination to those who have explored and thought about them.

Evolutionary changes associated with oceanic island biota

Once an organism becomes established on an isolated island dispersal ability ceases to be an advantage. The loss of dispersal ability among plants and animals is a widespread phenomenon. For example, birds of many different groups from penguins of the islands surrounding Antarctica to kiwis and moas in New Zealand, various ducks, and a wren on Stephen Island near the coast of New Zealand have become flightless. Perhaps Hawaii also had a flightless bird fauna before the wave of extinction that followed the arrival of the

Polynesians (Olson and James 1982). Hawaii has an exceptionally high proportion of flightless insects (Carlquist 1980). Plants also have their equivalent of flightlessness. For example, composites found on islands tend to develop strains that bear heavy seeds while related continental species have light and easily dispersed seeds (Carlquist 1965). The selection of low dispersal varieties on islands appears relatively straightforward. Animals and plants that migrate or disperse easily have a higher probability of leaving the island than those variants that are less mobile. Selection would then be for individuals that remain on the island.

Another characteristic of island organisms is their fearless or non-aggressive nature. The fearless nature of island organisms is attributed to disharmony, especially the lack of predators on islands. Accounts of the fearless nature of island birds to approach by humans abound (Carlquist 1965). In the case of birds, loss of flight and fearlessness may both be due to lack of predators. Most native plants of Hawaii have no poisons and other toxins : nor do they have well-developed thorns or spines, compared with their non-island ancestors. Hawaiian Bidens have poorly developed barbs on seeds, which could be in response to lack of predators or due to a lack of animals which in a continental setting are involved in dispersal. Island organisms lose mechanisms which increase survival ability under more predatory circumstances.

Island organisms tend to have lower rates of reproduction and mature more slowly than do related non-island populations. However, there are also reports of island "supertramp" bird species that can exploit open habitats such as islands disrupted by volcanoes or tropical storms and succeed as a group by building huge populations. These supertramp birds eventually become eliminated by species that more efficiently exploit the habitat, but they continue to maintain their species by moving on to other transient environments.

Many groups of organisms have shown a remarkable capacity to undergo changes in size and shape when they immigrate to isolated oceanic islands. This is a spectacular aspect of island organism, and many examples are found among both plants and animals. Plants that exhibit the herbaceous habit on continents have become trees on islands. This particular evolutionary change is referred to as arborescence. Of special note are a large number of trees found on St. Helena that evolved from a minimum of our different Compositae immigrants, three of which are closely related to asters.

Among animals, reptiles have been especially adept at filling various niches on islands that would usually be filled by mammals on continents. Of course, it must be noted that reptiles are also more capable of dispersal than terrestial mammals and as a result have had a greater chance to demonstrate their radiative capabilities on isolated islands than mammals have had in these situations.

Another example of unusual radiation is found among the Hawaiian land snails. As described above, native Hawaiian plants do not have the usual insect predators that continental plants have. Land snails have filled the niche normally occupied by insects, and they make up an extremely large and varied group of land organisms which have become specialized as leaf eaters. Speciation among snails has been favored by geographic isolation that has occurred on the islands as a result of the formation of deep valleys. Apparently the valleys have created impenetrable barriers between groups of snails, and in isolation the snails have undergone rapid speciation.

Many small islands have been reported as having extraordinarily high densities of spiders. The density is variable and the variations are explainable by the combination of distance to presumed sources of immigrants and the presence of bird or lizard predators (Schoener and Toft 1983).

The Hawaiian Islands are the most isolated on the Earth and provide excellent examples of the evolutionary process. Many studies have been published covering topics of natural history and evolution on Hawaii. The large island of Hawaii illustrates the evolutionary developments that have occurred.

in terms of dispersal there is no question that plants were the most able, while insects and mammals were the least able to immigrate to Hawaii. Of course, many groups were not even in contention as there are no native and mammals, reptiles. or amphibians. Many groups of plants are simply lacking. The evolutionary potential of Hawaii has been evaluated by dividing the number of endemic species by the number of likely immigrant species. The insects increased by a factor of 26, plants increased by 7, and the birds increased by at least 21½ times when the historically known birds are considered. On Hawaii as on other isolated islands, relatively few immigrants transformed the island into a luxuriant and magnificent assortment of plants and animals. What happens on islands in miniature has happened on the Earth on a grander scale. But on islands the process is most easily conceptualized and comprehended. The next section explores other kinds of islands and their contribution to an understanding of speciation.

Mountain tops, refugia, and other islands

Islands of plants and animals are not a feature only of the oceans They also exist on the continents and under water within the marine environment. In this sense, there are many kinds of isolated environments known as "habitat islands." These separated enclaves of organisms appear to be a prerequisite for genetic diversity and speciation. In this section, various types of biotic refugia and their significance are examined. A refugium is literally a place for refugees, often relict species. Biogeographically, a refugium is usually a place where a formerly abundant species persists even though its populations have declined or become extinct in the surrounding area. Generally the emphasis is an old species, but it appears that islands may be a refuge for new species as well. There are many refugia and a few prime examples will be described.

Nunataks

Ice age events were described along with the effect which these events had on the broad distribution of plants and animals. Within the glaciated areas of the continents, some isolated sites remained free of ice. These sites are called nunataks and functioned as refugia. In North America two sites are especially well known for ice-free conditions : the area of Wisconsin and southeast Minnesota called the "driftless area," and a series of nunataks which existed between the Cordilleran and Laurentide ice sheets. The Queen Charlotte Islands and other sites along the Pacific Northwest coast of North America were ice free and served as biotic refugia during the last glaciation (Warner et al. 1982). During colder glacial times these ice-free sites apparently enjoyed seasonal warmth and thus supported plants and animals which were able to expand their range as the glaciers melted.

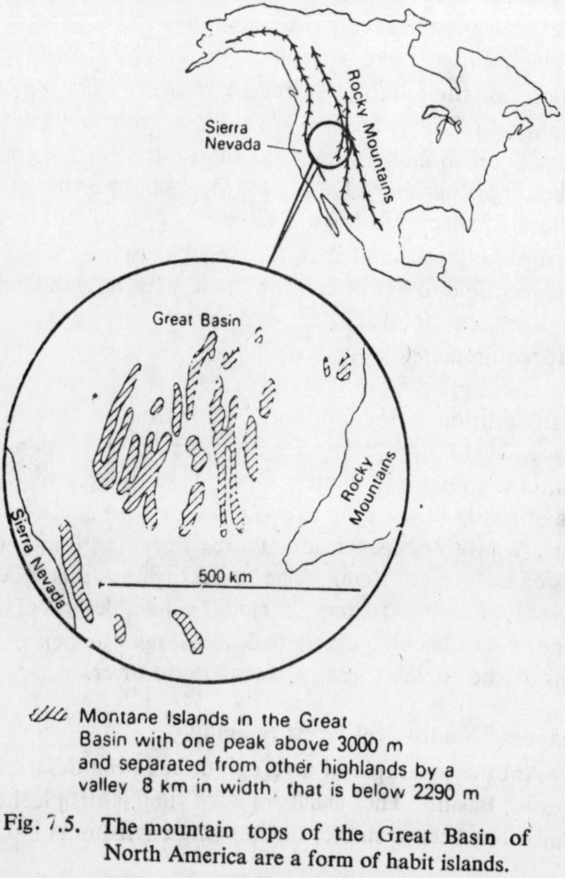

Fig. 7.5. The mountain tops of the Great Basin of North America are a form of habit islands.

Mountain tops and basin deserts

In contrast to nunataks of the past, there are also examples of refugia that exist in the present period of time. The mountains of the presently arid region of western North America have a flora and fauna which reflect a time when the whole region was colder and consequently wetter than now. Under current climatic conditions, the various highland habitats are isolated from each other by lowlands which are much warmer and drier. These mountains are thus biotic islands of cold-temperate and arctic flora and fauna separated from each other by a sea of grasslands and desert. As climates became warmer, the fauna and flora moved both in latitude and in altitude. The populations of organisms within mountain refugia have become increasingly vulnerable to extinction as their habitats became smaller. The areas of the mountain refugia vary in size from a few square kilometer to thousands of square kilometers, and in general the greater the area the larger number of species which exist on the mountains. Extinction of species follows a pattern related to population size, trophic level, and habitat specialization. The animals most susceptible to extinction are those with large bodies, which are high in the food chain, and have the most specialized habitat requirements.

In addition to the mountain refugium of western North America, there are many isolated basins that contain lakes which, in a cooler and wetter time, were interconnected as a series of rivers of a larger hydrological system. These isolated basins are now enclaves of an aquatic flora and fauna which are relicts of a former time. The desert pupfish is one of the survivors of the former period. Also, desert plants have expanded and become diversified in a large number of isolated basins as the climate became warmer and drier.

Pleistocene Neotropical forest refugia

Anotner example of a refugium has been described in the Amazon Basin. The usual view of the Neotropical rainforest is that it has been a stable, unchanging environment throughout the eons while higher latitude environments have undergone large fluctuations. Current evaluation of climatic and geologic change in the Neotropics reveals a far different scenario. During glacial episodes, the tropical rainforest was forced into retreat within isolated refugia while most of the Amazon Basin became a savanna or grassland habitat. The refugia may have become centers for speciation of tropical biota, and these centers have been identified by their large number of endemic species (Prance 1982). As described in the upcoming section on the evolutionary significance of islands, the latter view conflicts with more recent evidence concerning the high-latitude origin of speciation. Perhaps an alternative hypothesis would be that high rates of tropical speciation were initiated during less stable climatic and geologic conditions during the Pleistocene ice ages, and the new forms moved into the pockets of more stable environment, places which are now recognized as refugia. If this is the case then the tropical refugia were places which harbored the youngest species rather than relict species. When the climate became warmer and wetter, the new species expanded their range to become the basis for the development of the Neotropical rainforest.

Island refugia

The final example of refugia is that of oceanic and continental islands. The Juan Fernandez Islands and New Zealand provide examples of refugia of the ancient Gondwanaland biota of the Southern Hemisphere. Volcanic islands provide a haven to organisms that can reach them. The Juan Fernandez Islands lie west of South America and are a young volcanic island-seamount chain that have formed from a mantle hotspot as the Nazca Plate moved eastward. The islands are the habitat of ancient endemic taxa that have dispersed there and now survive refugially. In the case of continental islands, the biota which moves with the island is associated with the former continent to which it was connected. Many plant and animal species of New Zealand were apparently more widespread in the past when New Zealand was connected with Gondwanaland. One especially interesting example in the tuatara which is the only surviving member of a group of reptiles common during the

Mesozoic Era. This lizard-like animal does not have remarkable dispersal capabilities and survives in New Zealand as a relict of a former geologic era. Islands may thus contain biotic relicts which have no living continental relatives. As with other refugia these enclaves of organisms furnish variety to the genetic stock of the Earth's flora and fauna.

Evolutionary significance of islands

Islands have been critically important to the development of the science of evolution. Both Wallace and Darwin were inspired into thinking about the evolutionary nature of living organisms while studying the flora and fauna of islands. It is fitting that current ideas about evolution be considered in this chapter on islands.

Islands and habitat islands which have been described above have several characteristics in common. They are environments that withstand discernible physical stresses as a result of geologic and climatic change. They are places where species of plants and animals have become isolated fromrel ated species. The combination of geographic isolation of species and environmental stress appears to enhance the process of speciation and the development of novelties.

The theory of geographic speciation is an old idea which dates back to the German geologist, Leopold von Buch, who in 1825 laid out the theory. The modern advocacy for the theory stems from the work of Ernst Mayr and discoveries from work on populations of organisms. Speciation was once described as being a purely genetic phenomenon in which spontaneous or systemic mutations created new species. One of the major proponents of systemic mutations was Richard Goldschmidt (1940). Geographic speciation presents an alternative to mutation and proposes that speciation is a phenomenon of groups of organisms and their relationships to each other. That is, populations of organisms become splintered as a result of the dynamics of physical processes operating on the Earth. Speciation results when groups of organisms become isolated.

In this view, speciation comes about after groups of organisms become spatially isolated from each other. As described on plate tectonics, glaciation, climate, and rivers, there are many phenomena operating in different combinations of magnitude and frequency at various localities of the Earth. These act to isolate groups of organisms. Some areas of the Earth have relatively stable conditions over long periods of time while other areas are extremely dynamic and in constant flux and chaos. There is an interesting conflict of ideas concerning which of the environments of the Earth most actively promotes the major evolutionary development among organisms.

The traditional assumption has been that environments which support a large diversity of complex assemblages of organisms foster major evolutionary innovations. Under this assumption, the Tropics are opposed to the higher latitude terrestrial communities, and the offshore as opposed to the nearshore marine communities would be the wellspring of species innovation. The particular innovations of major interest are not simply variations, but the beginnings of distinctly different adaptations and forms which become distinguished as new genera, families, or higher taxonomic groups. The diversity of the Tropics and of the offshore communities alone may appear to demonstrate that these places are major sites of innovation. Indeed, the environment of the Neotropics, is considerably more geologically and climatically dynamic than usually assumed. Thus, traditional assumptions regarding both the static nature of the Tropics and the lack of innovation of stressful marine and high-latitude terrestrial environments, have not been borne out by research.

Evidence of origination of plant and animal species in the species-poor, cold Arctic latitudes After origination the animals expanded south into more favorable environments. A similar finding has also been described for the Southern Hemisphere. The paleontological evidence of recent plant and animal innovation supports the idea that the origins took place in higher latitudes where the oldest fossil remains exist, and then migrated toward the lower latitudes. Both hemispheres show the same pattern of innovation of life.

In the marine environment, research findings have been described for nearshore and offshore environments which provide further support to the idea that speciation is enhanced under conditions which are harsh and stressful and where only a few species are supported. In the nearshore environments there are fewer species, and those species which do exist have low rates of speciation as well as low rates of extinction. In contrast, species in the offshore communities are characterized by high speciation and high extinction rates. The overall consequence of slow species turnover nearshore and fast turnover offshore is that individual nearshore species are geologically older than are offshore species. However, when larger taxonomic groups or clades are examined instead of individual species, then the nearshore-offshore age distribution is reversed. That is, the nearshore species, which are generally older as individual species, are the most recently evolved forms of life, while the offshore species, which as individuals may be geologically young as a group, are the oldest forms of life. Thus, a transect across the seabed, from the shore to the edge of the continental shelf, is also a transect that passes from the most recent evolutionary inventions to the oldest forms of life.

Why do nearshore environments serve as the crucible of new forms of life ? The question has been approached in two ways. One possibility is that speciation results from the differential extinction rates of the nearshore and offshore communities. The nearshore communities have lower extinction rates. The greater extinction resistance of the nearshore species allows innovations to persist long enough to become diversified and then expand into the offshore communities. As an alternative approach, the differential of speciation within the nearshore and offshore environments may illuminate the contribution that geographic phenomena have on the process of speciation. The nearshore environment is a hazardous place for organisms to live in. Those few species which survive are extremely resistant to extinction and are generally distributed over wide geographic areas. The offshore environment tends to be relatively stable, and the species living in offshore communities tend to be restricted to small populations that are geographically confined to small areas. The difference in the two environments may enhance the origination of novel groups in the nearshore community. Because the species tend to be geographically widespread, some portion of the population is likely to be isolated in peripheral groups. These groups may have shift in developmental pattern which is sufficiently large to produce truly novel forms. This second explanation is similar to the geographic speciation theory discussed above with the addition of the idea that stressed conditions combined with species-poor environments from the crucibles of speciation of the Earth.

The reasoning of geographic speciation for terrestrial and coastal environments can be aptly applied to the processes of speciation on isolated islands. Islands are environments that often begin with a species-poor situation. Islands are subject to the same variety of high-stress conditions that affect large continents, but the effects of stresses and harsh conditions on organisms are exaggerated by the limited area of islands. They have demonstrated their propensity for providing fertile grounds for origination of species even more pointedly than marine and terrestrial communities. But there is an important caveat. Island floras and faunas although sources of novelty, do not appear to influence the evolutionary patterns of the Earth in the manner that sources of novelty in more continental locations or marine environments have done. Unlike continental and marine species, the novel origins of islands do not have a chance to expand into more stable environments as do the novel forms which develop in stressed environments of the continents and nearshore marine environments of the Earth.

Human disruption of islands

Humans have an enormous impact on the Earth's landcapes ; this is the principal topic of Part Four. Organisms which evolved within the isolated environments of ocean islands have been recognized as being especially vulnerable to human impact. Wallace was well aware that by the middle of the 19th century, island biota was already seriously diminished in quality when he stated : "in many cases it can never now be obtained owing to the reckless destruction of forests and with them of countless species of plants and animals". The present depauperate biota of many islands may be an artifact of human action rather than a function of naturally occurring processes.

Several excellent examples of human disruption of island biota can be cited from Hawaii. In Hawaii, the native plant species are 97% endemic. In the prehistoric past, the Polynesian cultures contributed many new plants to Hawaii and disturbed many natural habitats for agricultural purposes. They also introduced hogs, dogs, and chickens to the islands, which must surely have played havoc with the original biota. Additional changes accompanild Western contact. Europeans contributed to the destruction through the introduction of herbivores, especially goats and cattle, deforestation, and the introduction of a large variety of plants such as Koster's curse, which in the wake of deforestation have become extremely aggressive.

Serious debate took place between Egler and Carlquist regarding the fitness and capabilities of island species, Carlquist proposed that island species were genetically impoverished and lacked the ability to compete in a free context with "stronger" continental species. On the other hand, Egler argued that island species were as well adapted to their natural habitat as continental species are to their situations. He also pointed out that island species were poorly adapted to withstand grazing and predation and other forms of man-induced disturbance. He produced some good evidence that island ecosystems could withstand attacks from alien organisms provided certain features such as the forests were left undisturbed.

Given the nature of human impact on islands, no one can assume that the historically known flora and fauna of ocean islands are in a natural state of equilibrium. Equilibrium theory of island biogeography has been tested and, in the case of

Hawaiian birds, equilibrium by speciation rather than equilibrium by immigration has been determined to be congruent with the known data (Juvik & Austring 1979). New data further strengthen the idea of equilibrium by speciation. The birds of Hawaii illustrate the distinct possibility of reaching misleading conclusions because of unawareness of the effect that prehistoric human activities have had on island bird fauna (as Wallace had warned). At present the endemic land birds of Hawaii are represented by the following : one species each of goose, hawk, flightless rail, crow, flycatcher, and by two species of thrushes, five of honeyeaters, and 27 of finches or honeycreepers. Recent research on fossil remains of birds which became extinct before European contact but after Polynesian settlement on Hawaii includes at least the following additional endemic species : ten species of geese two flightless ibises, one sea eagle, one hawk, seven flightless rails, three species of owls (of a now extinct genus), two large crows, one honey-eater and at least 15 Hawaiian finches. There were at least twice as many colonizations and twice as many endemic birds as have been assumed from historically known fauna. There have been many critical discussions about the MacArthur Wilson equilibrium theory. The desire to apply the equilibrium theory or, for that matter, any theory to the solution of genuine problems fosters concerns about the adequacy of basic knowledge about patterns of life on the Earth and how there can be amelioration of the disruptive impact that humans have on other organisms.

Nature reserves and parks

There is an extremely practical side to the study of island biogeography. In the process of human expansion and settlement of the land, many species of plants and animals have been relegated to small parcels of land surrounded by large tracts of inhospitable clearings which have been prepared for agricultural uses. For example, in Europe and North America, immense areas of land were first cleared of their forests and converted into pastures and then plowed as agricultural fields. One of the outcomes was the establishment of diminishing patches of wooded lands (woodlots) surrounded by increasingly larger gaps as clearing proceeded. This process was well illustrated for Wisconsin farmland in the United States and for Europe. The modern event of deforestation of the Amazon Basin, at a rate of 11 million hectares per year, is unprecedented in the rate of change but not in the areal extent of habitat alteration. The uncleared lands become the reservoirs or habitat islands for the species which once occupied the whole area.

The establishment of nature reserves and parks as a means of protecting species from extinction is a major practical concern. Many issues about park planning seem to be akin to sland biographical problems. Can island biogeographical theory provide a basis for wise resource management ? Equilibrium theory poses some important unsolved problems. For example, it proposed that equilibrium becomes maintained through interaction among species. That is, the presence of one species is believed to change the probability for another species becoming extinct. Advocates of the model argue that if the number of species present is below the equilibrium level there will be higher probabilities of successful immigration and lower extinction rates. Conversely, high numbers of species will increase the extinction probabilities of all species and result in a decrease in the number of species.

In a study of avifauna on the English Farne Islands, the Welsh island of Skokholm, and a habitat island, Eastern Wood in Surrey, England, the means of species number regulation was tested. A Markov model, which treats species separately from each other, provided a means of simulating species number change. The results suggest that species richness is not determined by interactions between species, but rather by the probabilities of immigration and extinction which operate independently for each species. If interaction of species determines extinction rates of birds, then one large park would protect the most species. If species richness is determined by immigration and extinction probabilities which are a characteristic of each other may together provide higher probabilities for survival of avifauna than a single large park. Support for this conclusion came from the Nyika Plateau of Malawi where it was found that higher densities of birds were found on small forest patches because most species are adaptable forest-edge birds (Dowsett-Lemaire & Dowsett 1984). What best protects avifauna would most likely be unsuitable for other organisms such as large mammalian fauna. Thus, a variety of approaches including both large and small sites must be established to conserve species.

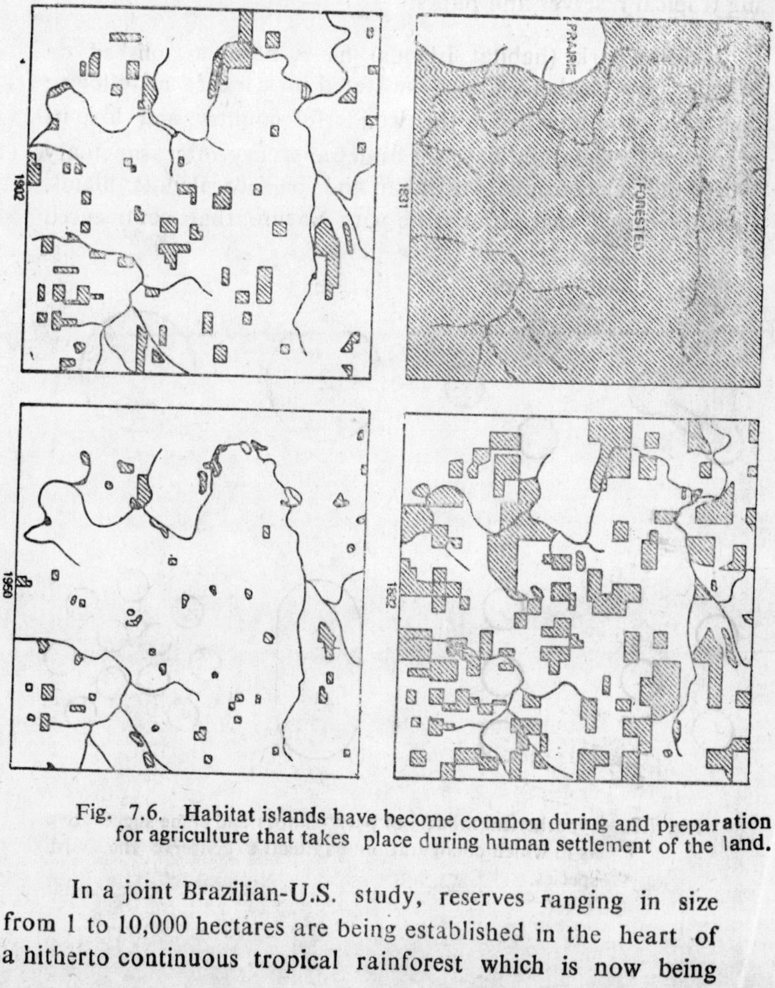

Fig. 7.6. Habitat islands have become common during and preparation for agriculture that takes place during human settlement of the land.

In a joint Brazilian-U.S. study, reserves ranging in size from 1 to 10,000 hectares are being established in the heart of a hitherto continuous tropical rainforest which is now being cleared. The reserves will be monitored for loss of plant and animal species during the next 30 years to determine the optimal size of a tropical part (Lewin 1984). Unfortunately, the results of the massive experiment will be available only after the major deforestation and habital disruption occurs. But the experiment should yield information on the rate of extermination of species which takes place as a part of deforestation and habitat island formation. It is hoped that the project will provide information needed to develop sound policies for creating tropical reserves and parks.

Some parks (habitat islands) have been established on islands. New Zealand, Cockayne stated his concern as follows :

We, who now live in this wonderful country, and love its marvellous vegetation, have set aside sanctuary after sanctuary where the palaeotropic, Australian and palaeozealandic plants, the survivors of that bitter strife with Nature, that commenced

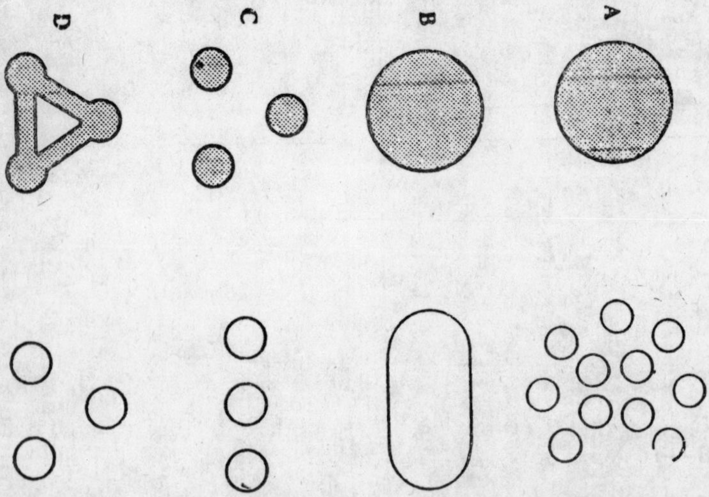

Fig. 7.7. Hypothetical models of park shapes and some suggestions as to which configuration may tend to conserve the most species.

millions of years ago, can still pursue their destinies if unmolested by their human enemies and the horde of foreign plants and animals he has let loose. Will our descendants prize this unique heritage from the dim past and preserve these sanctuaries intact ?

The situation of New Zealand is unique because naturalists were among the early wave of European colonists to arrive on the islands. They came in awe of the unusual flora and fauna of those lands and began early to set up parks and natural preserves to protect the native flora and fauna. The immigrants, however, also brought many species of plants and animals and attempted purposefully to introduce many plants and animals and thereby Europeanized the fauna and flora with the intention of "improving" the native biota. New Zealand originally had only two species of mammals, both bats. European immigrants introduced and established more than 50 species of mammals. The impact of one animal in particular, the red deer *Cervus elaphus*), has been described as profound and catastrophic because of its effect on the native forests and soils. A more recent assessment demonstrate the difficulty of determining the magnitude of effects of the introduced mammal. New Zealand has a long tradition of conservation through the establishment of parks and preserves.

The U.S. system of parks and reserves has been put together by a number of different institutions and organization in both the public and private sector of society. These lands vary in size from the enormous Alaskam wilderness areas to small plots. Most of the parks are habitat islands in areas that have been intensively used for hunting and trapping, logging, and farming, which disrupt fauna and flora to varying degrees. The United States has the advantage of a vast tract of land which encompasses a large variety of habitats within many climates.

The most exclusive Federal reserves are called Wilderness Areas and are part of an effort to set aside habitat for a variety of wildlife. There are also parks and reserves which have a variety of activities on them including lumbering, grazing, recreation, and hunting. There are state parks and reserves, and private holdings of land which have been purchased and maintained for various purposes. Some of the lands are privately held such as Nature Conservancy lands which in many cases are part of a system of small holdings designed to protect individual species. Other lands, such as wetlands, have been protected by citizen groups such as Ducks Unlimited and the National Wildlife Federation in an attempt to preserve particular "game" animals and to maintain their numbers for recreation purposes. Public and private efforts have been very important in the preservation of habitat.

There are many conflicting interests between government (especially military, energy, and transportation projects), mining, lumbering, recreation, hunting, and preservation groups, and these are played out in the public arena through political, legal, corporate, and citizen participation. Among the successful attempts to preserve island habitat is the work by conservation groups and the Royal Society of London that led to the abandonment of building a military air base on Aldabra atoll in the Indian Ocean.

The scheme would have covered an area now known to support 60% of the grant tortoise (*Testudo gigantea*) population. There are global efforts to preserve habitat by such groups as the United Nations and the Worldwide Fund for Nature. A global park system has been proposed, however, the complexities of habitat preservation are multiplied many times when countries are small, seriously overcrowded, desperate for food, and longing for economic improvement and industrial development. Yet it is poorest, especially tropical, countries which harbor perhaps 90% of the Earth's species. Habitat is rapidly being lost, to the detriment of many species. In an increasing number of cases the only hope for prevention of extinction of animal species is zoos (Holden 1984). The topic of human culture as an agent of habitat change will be taken up as part of the discussion of Part Four.

Conclusions

To biogeogrophers islands and their flora and fauna have been essential to the acquisition of basic knowledge about patterns of life on the Earth. The following points summarize major themes of the present chapter. Species of plants and animals have varying abilities to disperse. As a result, oceanic island communities tend to consist of highly selected groups of organisms which either survive long-distance travel or were able to with-stand the rigors of movement as conditions permitted.

For biographers the role that geographic characteristics (area, distance to nearest continent, distance to other islands, whether continental or oceanic in origin, ocean current, climate, etc.) play in the plant composition of a particular island is of special interest. Many examples of filtering, impoverishment, and disharmony illustrate way that distance affects dispersal. However, while distance may be important to present-day distributions of island biota, distance alone is often a very predictor of biotic affinities between islands and their nearest continents. This fact was described and puzzled over by Wallace and Darwin more than a century ago. Geologic and climatic history of an island often yield clearer understanding of biotic affinities than distance. This does not preclude distance as a factor; rather, it means thatd istance and physical and biotic events in the geological past are all part of the explanation of patterns of life of modern island.

Much current knowledge about speciation has been stimulated by the study of islands and their fascinating flora and fauna. Islands are not merely receivers of immigrating organisms, they are crucibles of organic change. They clearly demonstrate that geographic isolation alone is sufficient cause for speciation. If two groups of individuals of the same species are separated for a long enough period of time, they tend to change enough to be recognized as separate species. Islands support many examples of organisms that have radiated to fill niches that are unusual for their continental counterparts. They illustrate the plasticity of organisms and their capacity to adjust and adopt to various physical and biotic conditions when given the chance.

The continental counterpart to islands in the sea are various habitat islands such as nunataks, mountain tops, and inland basins. These serve as refugia and tend to harbor unusual biota compared to the surrounding area. Species-poor and stressful environments. These environments occur as part of the long-term geologic and climatic variation that takes place on continents. For example, ice-free areas become isolated when glaciers build up around their periphery and movement of organisms becomes restricted for long periods of time. When conditions change to enhance movement of individuals, novel organisms that evolved in the stressful and species poor environments are then able to expand their range into more favorable environments. Islands of the sea and habitat islands are thus similar because they are both isolated places which have enhanced rates of speciation and harbor unusual biota. In this way they contribute to the biotic richness of the Earth.

Islands provide an experimental cornerstone to biogeographers in their attempt to understand the patterns of life on the Earth. In the past they stimulated thought about the relationships between organisms and their environment. In the future islands, especially habitat island such as parks and reserves, will likely play an important role in maintaining biotic richness. Islands have influenced and will continue to affect the Earth's biota to an extent that is far out of proportion to their geographic area.

8

Active Animals of the Deep-Sea Floor

BY JOHN D. ISAACS AND RICHARD A. SCHWARTZLOSE
OCTOBER 1975

Baited automotic camera dropped to the bottom of the ocea nreveal a surprising population of large fishes and other scavengers that find and consume dead animals that fall from the water for above.

Photographs made by automatic cameras that have been dropped to the deep-ocean floor have confirmed a finding that was first suggested by evidence gathered from baited traps some years ago. A deep-sea population of large, active animals thrives in what was generally assumed to be a province inhabited mainly by small, feeble creatures such as worms, snails sponges. The thousands of pictures make it clear that much of the deep-sea floor teems with numerous species of scavengers : vigorous invertebrates and fishes, including some gigantic sharks, that are supported by a marine food web whose extent and complexity is only beginning to be perceived.

The celebrated expedition of H.M.S. *Challenger* in the 1870's laid to rest the old idea that the deep waters of the open ocean were a lifeless desert. The *Challenger's* trawls and dredges brought to the surface and to the attention of biologists a vast collection of deep-mid-water and bottom-dwelling creatures from even the deepest ocean trenches. The inventory included some of the most some of the most grotesque forms into which higher animals have ever been molded by adaptation to extreme conditions, andseveral animals that had been thought to be long extinct.

In this contury much has been added to that inventory and to our understanding of deep-living animals. The investigations were conducted, however, by means of trawls and dredges, by direct observation from deep-diving submersible vehicles and by inspection of photographs made by cameras lowered on cables from ships, and it is characteristic of these methods that they ordinarily sample only the stationary or slow-moving animals of the deep-sea communities; the deeper the investigation is, the more difficult and the less effective it is. The gave rise to the prevalent assumption of a few years ago that the ocean bottom was sparsely inhabited by weak creatures specially adapted to live on the only food material then thought to be available at great depths: a terminate food web supported by the thin but constant rain of detritus that sifts down from the surface layers and that is metabolized by primitive filter feeders or bacteria and deposit feeders in the bottom ooze. A proper reconnaissance of the active deep-sea creatures whose presence was revealed by the baited traps required new tools.

Over the past seven years Meredith Sessions, Richard Shutts and the authors, working at the Scripps Institution of Oceanography, have designed and constructed robot motion-picture and still cameras and other instruments with which to explore the bottom and to study the nature, distribution and behavior of its inhabitants. We find, first of all, that the population of sea-floor invertebrates is by no means sparse and that many of its species are far from weak. And they are not alone. Their domain is shared by a population of scavenging fishes and crustaceans adapted for the prompt discovery and consumption of large falls of food ; the bodies of dead animals descending from above, mid-water creatures that happen to approach the bottom and juveniles of their own species that return to the deeps from the shallower water where they undergo their early development.

The cameras we have devised for the ocean-floor study are free-fall devices connected to a recovery buoy, a flood-light and a bait holder. Released at the surface, the camera falls to the bottom and remains there, anchored by the bait holder for between 12 and 48 hours, making still photographs or short bursts of motion pictures at five-to 15-minute intervals. The bait is in the foreground of each frame and the camera's view is either vertical or oblique. At the end of the mission the bait ballast is released and the camera and its buoy rise to the surface, where a radio transmitter broadcasts a signal that aids in recovery. The free-fall technique allows a research ship to conduct various other missions while a number of cameras it has distributed keep functioning on the sea floor. And the work can be done from rather simple, inexpensive vessels, provided only that they have reasonable navigation and sonic-sounding equipment.

In a typical drop the camera is released in water that is between 400 and 7,000 meters deep. Greater depths call for a special camera housing and flotation gear because of the immense pressures. The camera operates during the descent, which may take as long as several hours, but ordinarily not much is seen in photographs made on the way down except for the "snow" of small particles, which are ubiquitous in the mid-waters of the sea, and a few fleeting small crustaceans.

Sometimes the very first photographs on the bottom, made a few minutes after the camera reaches it, show great activity, with fishes and invertebrates already tearning at the bait. More commonly, however, the first scenes show only brittle stars, large shrimps, amphipods (small crustaceans) and perhaps a fish or two. On more than half of the missions at least one fish has been photographed within 30 minutes of the camera's arrival on the bottom, even at the greater depths. (Curiously the first photographs sometimes record a fish of a species that is not seen again in the sequence, as though the camera had invaded the territory of a creature that was not a part of the population of active scavengers and wanted nothing to do with the subsequent activity.)

Usually the number of fish gathered around the bait increases slowly, reaching a maximum after a few hours. Often the scene develops into one of furious activity, with several species of fish competing for the bait, thrashing and tearing at it and sometimes attacking one another. Shrimps, brittle stars, amphipods and other invertebrates encroach on the melee. In almost half of the sequences from drops down to 2,000 meters the party ends abruptly after three to eight hours, when some creature, usually a large shark, moves in, frightens off the other fish and consumes the bulk of the bait. In any case the time comes when most of the bait has been eaten. The fish depart, and slowly crabs, sea urchins, snails and other such creatures arrive to complete the task of sanitizing the sea floor.

Sometimes the direction of the nearbottom currents has been determined and it can be seen that these latecomers to the banquet plot upstream toward the bait-an indication that they are probably following a scent. The fishes also probably depend on scent for close-in detection of the bait; on some occasions we could relate the number of fish gathered around the bait to the strength of the current, which suggests that an increased current had carried the scent more widely. Like wide-ranging scavengers on land, however, fish that are farther away are probably led to the area of the bait by other cues. They may sense the successive collapse of loosely established territories as the scavengers that held them move closer to the bait, each invading the area once held by an absent neighbor. The western prairie wolf, vultures and other terrestrial scavengers respond to just such a territorial collapse to converge on large kills.

Surprisingly, we find particularly abundant bottom populations in regions of the Pacific where the surface populations are the least abundant, and vice versa. For example, under the least productive surface waters of the North Pacific Gyre, on an austero bottom paved with manganese nodules nearly 6,000 meters down, more than 40 large fish and shrimps were attracted by the bait within a few hours; at least four species were represented. Moderately productive areas of the Indian Ocean, the California Current system and biologically poor areas around the Hawaiian Islands have also shown remarkably large numbers of deep-sea creatures. On the other hand, only 4,000 meters down under the most productive oceanic waters of the world, in the Antarctic over a bottom of soft organic ooze, the bait was visited by only a few eelpouts, brotulids and rattail fish (grenadiers), along with some small crustaceans, the louse-like isopods. The brotulids (fishes distantly related to cods) remained almost motionless for hours, nibbling gently at the bait. Camera drops along the underwater ridges of the Line Islands, a highly productive equatorial area of the Pacific, revealed only a few eels. Photographs from thc bottom below the rich Peru Current showed very few fish but large numbers of invertebrates : furious masses of amphipods that stripped the bait in a few hours.

The explanation for such a puzzling distribution must lie in the nature of this part of the marine food web. These abyssal roving scavengers must depend for their sustenance in substantial part on falls of large food particles. Such particles include fish and marine mammals that have died, fragments of forage from surface-feeding schools of fish, vertically migrating creatures, garbage from ships and (nowadays) animals killed by under-water explosions, ships' propellers and whaling. Falling food material can support a population of deep scavengers only if the food descends to the bottom, and it may be that the mid-water population below highly productive regions is sufficiently dense and continuous to consume such food on the way down, Regions of low productivity, on the other hand, may support such sparse and discontinuous mid-water populations that a substantial proportion of dead surface creatures do fall all the way to the sea floor. And there on the floor even a meager fall can support a sizable population, whose scavenging can be much more intensive than that of fish living on the relatively brief passage of falls through the immense volume of the mid-waters.

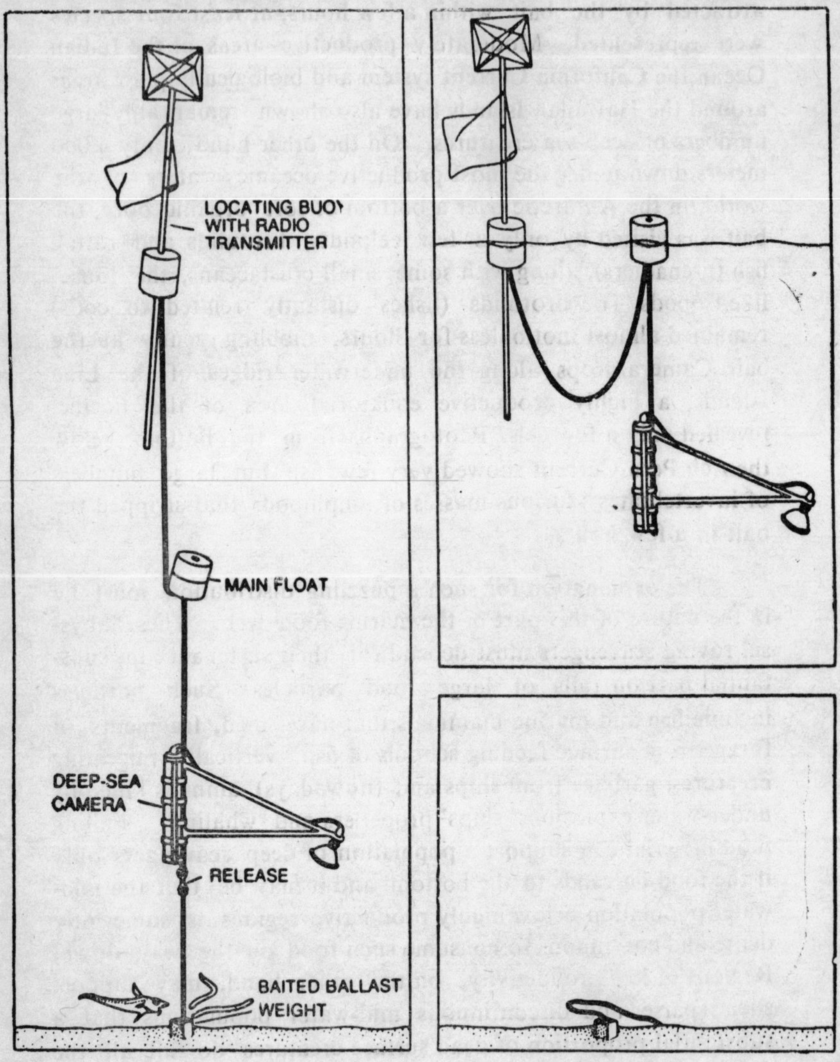

Fig. 8,1. Free-fall Camera is dropped to the ocean bottom, where it is held in position by a baited ballast weight and a float made of foamed plastic.

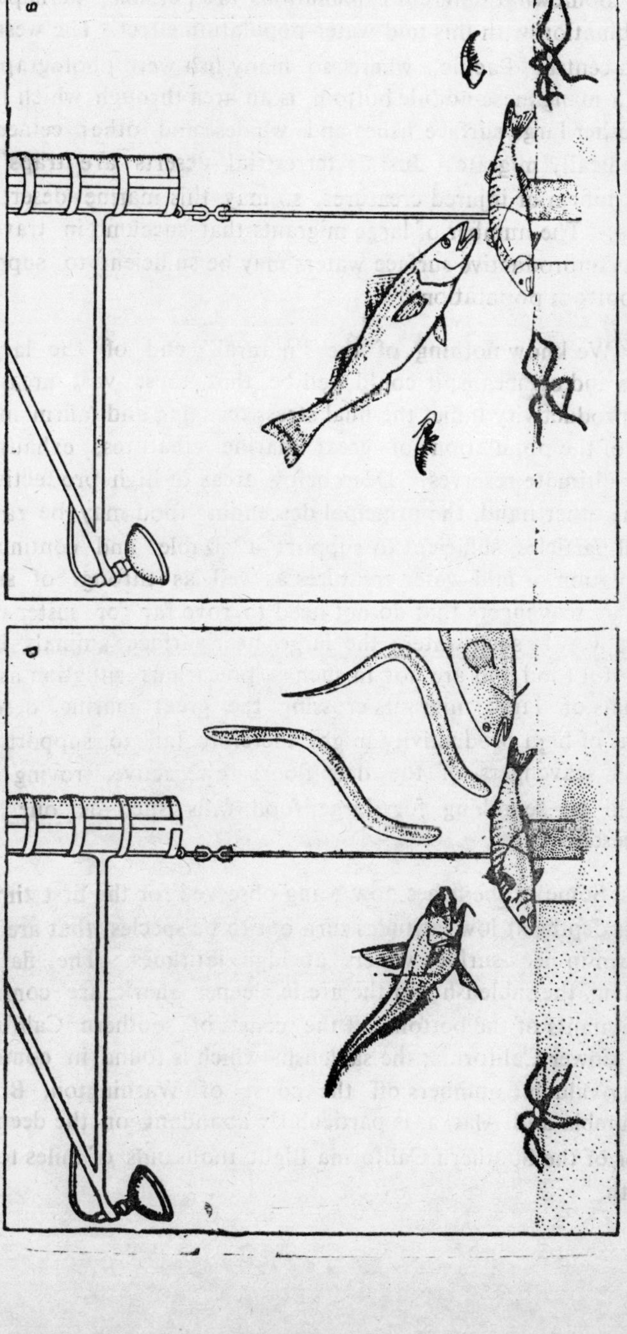

Fig. 8.2 One likely sequence of events following the arrival of a baited camera system on the ocean floor is shown in the idealized drawings on these two pages.

Somewhat different explanations are possible, perhaps in combination with this mid-water-population effect. The western north-central Pacific, where so many fish were photographed over a manganese-nodule bottom, is an area through which tuna and other large surface fishes and whales and other cetaceans periodically migrate. Just as terrestrial deserts are traps for aged, infirm or injured creatures, so may this marine desert be a trap. The number of large migrants that succumb in traversing its unproductive surface waters may be sufficient to support the bottom population.

We know nothing of the "natural" end of the largest fishes and cetaceans; it could well be that these vast areas of low productivity inflict the final stress on aging and infirm members of the populations of great marine creatures, exhausting their ultimate reserves. Deep below areas of high productivity, on the other hand, the principal descending food may be rather small particles, sufficient to support a sizable and continuous population of mid-water creatures as well as throngs of small bottom scavengers that do not need to rove far for sustenance. Moreover, in such waters the large near-surface animals have plentiful food and are not in such a precarious situation as the schools of large migrants crossing the great marine deserts. Areas of high productivity might therefore fail to support the larger scavengers of the deep-floor; few active, roving fish would be searching for large food falls such as our bait represents.

Some of the fishes now being observed for the first time at great depths at low latitudes turn out to be species that are well known in near-surface waters at high latitudes. The flatnose codling, the sablefish and the arctic sleeper shark are common inhabitants of the bottom off the coast of southern California and Lower California; the sablefish, which is found in commercially valuable numbers off the coast of Washington, British Columbia and Alaska, is particularly abundant on the deep-sea floor of the Southern California Bight thousands of miles to the south.

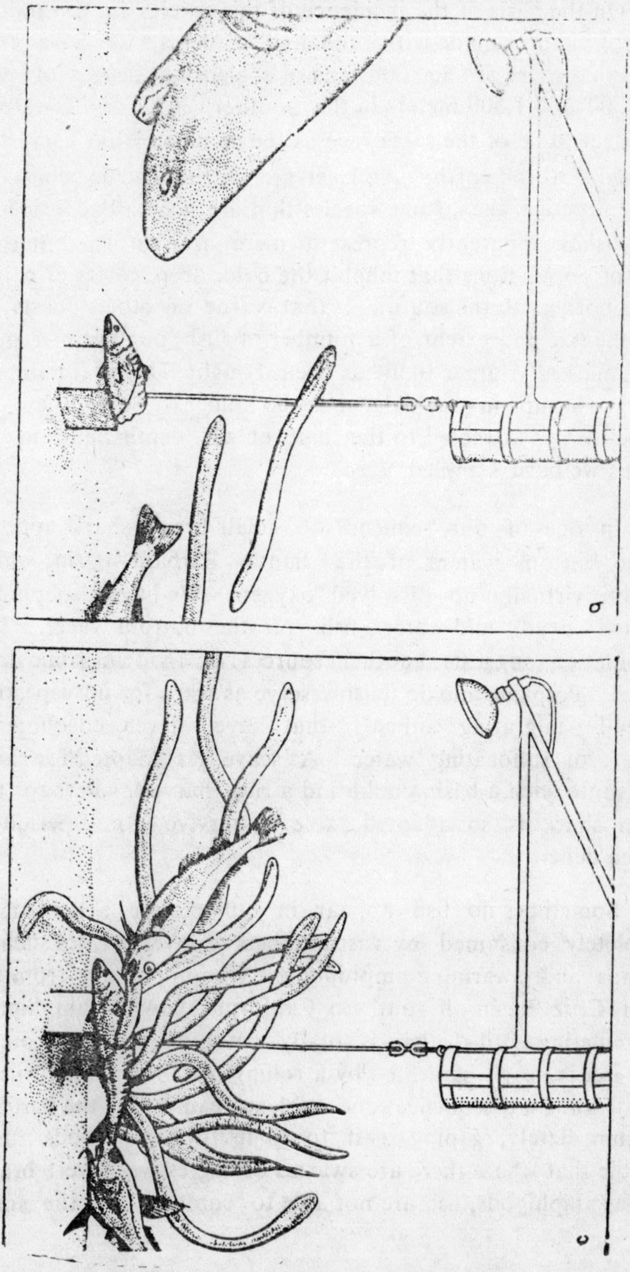

Fig. 8.3. Bait increases slowly, reaching a maximum after a few hours.

On the basis of the incidence of this species in a random series of pictures made with unbaited comeras, we have estimated that there are 800,000 tons of sablefish at depths of between 800 and 1,500 meters in the southern California waters; they seem to be of the same race as the commercially harvested variety far to the north. At least along the Pacific coast of North America, then, some species that are generally considered arctic fishes apparently represent mere near-surface outcroppings of populations that inhabit the cold, deep waters of continental borders to the south. If that is true of other coasts as well, the size and extent of a number of fish populations may be dramatically larger than has been thought. The total numbers cf large shrimps in particular must be immense, since they are almost always attracted to the bait at all depths and in all regions we have sampled.

In one of our sequences a small brown shark appears in the bottom waters of the Santa Barbara Basin, which contain virtually no dissolved oxygen. We have also photographed dead mid-water fish on the bottom there. The combination suggests another sources of food in some deep waters. Perpaps anoxic basins serve as traps for unsuspecting vertically migrating animals that have never encountered pockets of suffocating water. A scavenger adapted for short forays into such a basin could find a rich harvest. Perhaps the brown shark is so adapted, able to survive a brief period of oxygen deficit.

Sometimes no fish appear in our pictures : the bait is completely consumed by vast numbers of invertebrates such as shrimps and swarming amphipods. Motion pictures from the Santa Cruz Basin off southern California show the amphipods accumulating until the bait is totally covered and the surrounding water is nearly saturated by a roiling mass of these crustaceans. In that sequence one sablefish came near the bait but left immediately, gaping as if to rid itself of amphipods. Is it possible that where there are swarms of aggressive invertebrates such as amphipods, fish are not able to compete for the small amount of food that reaches the bottom? Amphipods can sustain themselves on small food particles, but they are also capable of quickly and completely devouring the much larger pieces that fall to the ocean floor. Deep-living octopuses are other invertebrates that tend to monopolize the bait. In one motion-picture sequence made at about 4,000 metres a small octopus squats on the bait, keeping the grenadiers at tentacle's length, and in an unusual sequence made with a still camera off Cedross Island in Lower California two octopuses fend off a single grenadier from the bait.

The bait is often taken over aggressively and completely innumerable hagfish, primitive eyeless and jawless chordates that dril into dead creatures and consume them from the inside out [see "The Hagfish," by David Jensen : **Scientific American** Offprint 1035]. Hagfish thrive at a depth of from 200 meters to nearly 2,000 meters. At first we were puzzled by the reluctance the Gorgon's-head tangle of hagfish and feed on the bait. Closeup motion pictures gave the answer : the hagfish enclose the bait in a thick cocoon of slime that other fishes apparently find distressing. On a number of occasions fish emerge from the feeding mass making frantic efforts to clear their gills of slime. The spectacular ability of the hagfish to exude slime has long been known ; the exudation of a single hagfish can convert a large container of water into a slimy gel. Their employment of this defense mechanism to sequester food, however, had not been suggested.

Some of the scavenging bottom fishes display a feeding capability that could scarcely be predicted from an examination of their jaw and tooth structure. The grenadiers are quite capable of tearing out the abdomen of a bait fish ; sablefish shake large baits as terriers do and spin furiously to twist off mouthfuls of food.

The large sharks that frequent the deep-ocean floor have been photographed to depths of about 2,000 meters. Their behavior in approaching the bait appears to be mediated in part by a sense of smell : the fish execute slow, deliberate geometrical maneuvers that apply combine to establish a complex a pattern. On some occasions, when bait holder has been hung somewhat as a rough bottom, the sharks are unable to discover it. Clearly they accustomed only to food resting on sea floor ; when they connot find it, snudge and bite at rocks or other por objects under the bait. Even search for bait that is on the ocean usually requires more than one before the bait is found. Several are sequences end just as the shark position to seize the bait, with the dead ahead within the width of park's jaws. Yet the next sequence, five minutes later, may show the untouched and the shark still end in a slow, deliberate search. Like as. hagfish are unable to discover the when it is a meter or so above bottom, whereas eels and grenadiers no difficulty locating bait that is above the bottom. Knowledge of feeding limitations will help in the development of better techniques for some deep-water fisheries, where the depredations of shark and hagfish greatly limit the catch.

One of the shyest fishes we have photographed, and perhaps one of the deepest-living, is the brotulid. In a number of photographs from the deepest locations brotulids lurk at the outer edge of a group of grenadiers. In one motion picture series a brotulid hovers like a motionless blimp through sequence after sequence, facing the feeding grenadiers. Unlike most of the deep-water fishes, the brotulids have very reduced eyes, and it may be that their behavior is related to diminished visual acuity. Creatures living more than several hundred meters below the surface are maneuvering in a profound darkness that is lighted only by the faint glow or brighter brief flashes of bioluminescent. Only fishes with a highly developed visual system can be expected to be guided by visual cues. Brotulids may be responding not to the presence of the bait but to the sound of the feeding fish.

Many sequences yield fresh insights into feeding behavior. A crab gingerly lifts some sea urchins off the bait, holds them away from its body like a spider ejecting a distasteful insect from its web and drops them. Sheltered by the empty bait holder, a small spiny lobster flails its antennae in a strong current, apparently grasps a small swimming crustacean between the antennae and conveys it to its mouth by some movement too quick for us to make out. As grenadier goes after small food particles in the sediment with a sudden explosive thrust into the bottom, throwing a cloud of sediment through its gills.

The uniformly large size of most of the fish photographed at great depths presents something of a puzzle. Very few small fish are ever seen. The grenadiers photographed at from 750 to 6,000 meters in the Pacific, Antarctic and Indian oceans are all large, nature fish, some measuring more than a meter in length. We have so far recovered no free grenadier eggs or juvenile fish from collections made off California anywhere between the surface and the bottom, although females with ripe ovaries have been collected near the bottom. All the arctic sleeper shark photographed off souther California and Lower California have been very large, but most of the photographs have shown only a small portion of their total length, which we can only estimate as being between five and eight meters. They also must be quite common, since they have been photographed in nearly half of our missions off California down to 2,000 meters ; on a number of occasions when more than one large shark was photographed during a mission, we could tell from distinguishing scars and other markings that the sharks were different individuals.

It is probable that the juveniles of these deep-living fishes inhabit much shallower depths than the adults. The young of the sablefish, for example, are numerous in many places at depths of 100 meters or so. It may be that the rarity of juveniles is merely the result of great adult longevity and low fecundity. On the other hand, juveniles of many species must return to the deep bottom environment, indeed, their return may constitute a meaningful importation of food fromt he more productive upper layers for the nourishment of the total adult population. We usually think of the relation between juveniles and adult stocks only in terms of replacement. Juveniles may also be important as prey ; indeed, in some freshwater environments the young of a species are a principal prey of adults of the same species. To the degree that this process, which has been called Faginism, is important to the food economy of the deep-sea populations, we would of course expect to observe a paucity of juveniles in the populations : most of them are quickly consumed by their elders.

The baited camera suffers from some limitations common to many simple exploratory tools. Its sample is highly selective. Quantification and even identification can be doubtful. There are several ways to deal with these problems. One can, for example, set out fishlines and traps in order to retrieve specimens for sure identification. A particularly promising means of quantification is an unbaited drifting camera that lightens itself until it rises just a few meters above the bottom. It drifts with the current, making overlapping photographs and meanwhile recording the direction and distance of drift. Such a camera has been successfully operated in waters about 1,000 meters deep. We have plans to develop a drifting camera that will remain submerged for several months, making photographs along a drift track as long as 200 kilometers. The drifting camera should yield a meaningful census of the creatures that have been observed by stationary baited cameras and thus help us to assess the potential food harvest from the deep-ocean floor and to understand the ramifications of this remarkable branch of the marine food web.

Meanwhile vast areas await investigation with the baited camera. A series of photographic "section" across the continental shelves along all the major land masses and down the slopes to the abyss would fill many gaps in our knowledge of bottom-dwelling animals and should reveal entirely new fishery resources. Among the environments that are of special interest to oceanographers and marine biologists are the floor of the Arctic. Ocean, the deep delta of the Congo River, the Antarctic slope, the deeps of the Mediterranean, the top of seamounts and mid-ocean ridges and the slope and bottom of deep occanic trenches. Since the free-all cameras can be operated from inexpensive, unspecialized craft, they are particularly suitable for studies by investigators in underdeveloped countries who want to know more about the creatures and the potential deep fisheries off their coast.